American Institute of Chemical Engineers

Topical Conference on Novel Technologies 2007

Held at the 2007 AIChE Spring National Meeting

April 22-27, 2007
Houston, Texas, USA

Printed from e-media with permission by:

Curran Associates, Inc.
57 Morehouse Lane
Red Hook, NY 12571
www.proceedings.com

ISBN: 978-1-60423-361-2

Some format issues inherent in the e-media version may also appear in this print version.

ISBN: 978-1-60423-361-2

American Institute of Chemical Engineers
Proceedings
Three Park Avenue
New York, NY 10016-5991
Phone: 212-591-8100

www.aiche.org

American Institute of Chemical Engineers

Topical Conference on Novel Technologies 2007

Held at the 2007 AIChE Spring National Meeting

AIChE

April 22-27, 2007
Houston, Texas, USA

American Institute of Chemical Engineers

Topical Conference on Novel Technologies
2007

TABLE OF CONTENTS

Session 34 – Novel Reaction Technologies

Session 55 – Novel Separations Technologies

Session 75 – Novel Enabling Technologies

Session 96 – Miscellaneous Novel Technologies

Experimental Studies of Flow in Trickle Beds

Timothy J. O'Hern[1], James E. Miller[2], Walt Gill[1], and Lindsey Evans[3]. (1) Engineering Sciences, Sandia National Laboratories, 1515 Eubank SE, MS0834, Albuquerque, NM 87123, (2) Sandia National lab, 1001 University Blvd, SE, suite 100, Albuquerque, NM 87106, (3) Sandia National Labs, 1001 University Blvd, SE, suite 100, Albuquerque, NM 87106

An extensive series of experiments on flow regimes in gas-liquid cocurrent downward flow through a variety of fixed bed materials has been performed. Controlled variables included gas and liquid flow rates, bed porosity, and liquid viscosity. The primary measurement was pressure at various axial locations along the bed. Based on the results of these tests, a larger test apparatus was designed and built to allow measurement of interfacial area between two immiscible liquids in a gas-liquid-liquid flow. The test loop, which was designed to provide an experimental means of varying important parameters while observing reaction rates, has been brought on line. Full independent control of both liquid and gas flow rates has been demonstrated. Loop component flow characteristics have been obtained, and a preliminary operational model has been developed for setting up future experiments. The interfacial areas will be measured by following the progress of an interfacial reaction with known kinetic parameters. The Lewis cell technique has been applied to determine kinetic parameters for the base-catalyzed hydrolysis reaction of fluorescein diacetate occurring at a water/organic interface of known area. Modified Lewis cell techniques were developed and implemented to increase the rate of data collection.

Modular Microreaction System - a Powerful Platform for Product and Process Development

Shaibal Roy[1], Olaf Stange[2], Matthias Kroschel[2], Frank Herbstritt[2], Karoly Nagy[2], Thomas Daszkowski[3], and Eric Boonstra[3]. (1) Process Technology Group, Bayer Technology Services, Americas, 8500 West Bay Rd; MS #52, Seabrook, TX 77586, (2) Ehrfeld Mikrotechnik BTS GmbH, Mikroforum Ring 1, 55234 Wendelsheim, Germany, (3) Bayer Corporate and Business Services, 8500 West Bay Road, Baytown, TX 77520

Microprocess technology opens new pathways for the development of economical and innovative processes. In the chemical and pharmaceutical industry, this technology plays an increasingly important role in research, development and production because of its well known decisive advantages over conventional technology.

The Modular Microreaction System offers a powerful development and production tool combining these advantages with great flexibility and a wide scope of application for process development and intensification. The system comprises individual microstructure-equipped modules that can be freely arranged on a solid base plate, enabling the fast and flexible setup of a wide variety of microreaction plants. Tensioning modules hold the individual modules tightly together.

Some modules perform basic process operations such as mixing, dispersing, heat transfer, etc. More specific modules meet the demands of low temperature reactions (e. g. -80 °C), precipitation reactions (e. g. catalysts, pigments) or heterogeneous catalysis. Finally, the Modular Microreaction System can be used for liquid phase, gas phase and multiphase reactions in heterogeneous gas-liquid systems. Standard operating parameters include a temperature range of –20 to 200 °C; operating pressures up to 100 bar (at room temperature); and rated throughputs of up to 25 L/h. Special modules available significantly extend this process range.

The Modular Microreaction System provides the user with a flexible, safe and costeffective platform for the development of new synthesis methods for the production of chemical compounds and improving existing processes through intensification. Continuous operation and the more efficient process control made possible by microprocess technology often result in significantly shorter development times and thus reduced costs. Each plant configuration is tailored to the requirements of the reaction, and can be extended at any time with additional modules and components. New modules and applications will be shown.

A New Approach to Nanotube Based Bulky and Thin Film Material: the Composites from Polyaniline and Derivatives

Mian Jiang, Department of Natural Sciences, University of Houston – Downtown, One Main Street, Houston, TX 77002

Nanotube based composite material has gained wide attention recently due to its unique thermal, electrical, and mechanical properties. Most of these preparation involve chemical synthesis, chemical vapor deposition, plasma fabrication, and most often, result in insulating composites. Conductive nanotube composites, on the other hand, usually use approaches such as layer-by-layer assembling, guest-host chemistry, and template imprinting techniques. These techniques often generate non-homogeneous composites that are subject to unexpected features and unpredictable properties.

In this work we present a new approach to make nanotube based bulky and thin film composites. Our method employs a high energy curing treatment of the nanotube-aniline dispersion, followed by a solution electrochemical conditioning. The curing process can be any energy-elevated treatment like UV radiation, heating, or others, whereas the subsequent electrochemical conditioning can be carried out in aqueous media with wide pH range from extremely acidic to weak basic conditions. The model monomers for our proof-of-concept experiments were aniline and multiwalled carbon nanotubes (MWCNT).

The resultant composite can be formed in either bulky or surface membrane state, depending on the reaction media and support matrix. In our study, we focused on the surface membrane protocol because of its ease for subsequent surface characterization. This formed polyaniline (PAn) – MWCNT composite film is stable, redox-active, and thickness-controllable. Moreover, our new approach for fabrication produces the more homogeneous composite than the composite generated with procedures described previously. For the aspect of potential application, the new PAn-MWCNT exhibited electrocatalysis, high capacitance, and strong adsorption properties. Although our discussions have been focusing on the PAn-MWCNT surface membrane, our finding also revealed single walled carbon nanotubes (SWCNT) and other aniline derivatives have similar features as PAn-SWCNTs. A more systematic study is underway.

This work was supported by the ORC grant from UH-Downtown, the Starter Award from Society of Analytical Chemists, the US ARO grant (No. W911NF-04-1-0024), and Welch Foundation (No. BJ-0027).

Finding New Catalysts and Process for the Production of Acrylic Acid from Propane

Scott Han[1], Abraham Benderly[1], and Wolfgang Ruettinger[2]. (1) Monomers Research, Rohm and Haas Company, 727 Norristown Road, Spring House, PA 19454, (2) BASF Catalysts LLC, 101 Wood Avenue, Iselin, NJ 08830

In our process of evaluating several different pathways from propane to acrylic acid, our approach was two-pronged: 1) Evaluate processes for technical feasibility and 2) Evaluate and develop catalysts which can be used in these processes. The constraints we had were to develop a process with favorable CO2 production balance versus existing processes (green chemistry) and to be economically favorable. We developed several alternative processes which were evaluated from a process perspective versus the known two-step production of propylene from propane and subsequent oxidation of propylene to acrylic acid. The current favorite process uses CO2 as an oxidant in addition to oxygen to form propylene (ODH reaction), followed by conventional oxidation of propylene to acrylic acid. This process possesses a favorable energy balance and high propane conversion potential. Catalysts for the involved ODH reaction with CO2 and oxygen as oxidants were found and are currently being optimized.

H.E.N.C.I. DISPERSED-PARTICLE BED TECHNOLOGY AND THE ADVENT OF HIGH MASS- AND MOMENTUM- TRANSPORT EFFICIENCY NANOCATALYTIC UNIT OPERATIONS

When employed in-situ, many new nano- (and micro-) sized catalyst particles/structures have been shown to effect extremely favorable (rapid) extrinsic kinetics[1,4], largely due to their great surface-area/mass ratios. This effect is extreme in catalytic scenarios in which the rate-limiting step(s) include mass transport to/from the catalytic surfaces (such as a packed-bed reactor being fed with an ultra-dilute-solute reactant), and has extremely positive economic ramifications for applications requiring high-throughput, high-conversion *and* high-mass- (& momentum-, & often heat-) transport efficiencies to achieve cost-effective operation (hereafter acronymed **HTCT3** applications). Rapid overall kinetics hold the potential to greatly increase process efficiencies of niche catalytic CPI processes as well, and the desire to use these highly effective catalysts for various CPI *as well as large-scale GWR applications* has been a natural consequence. Catalytic nanoparticles themselves, *often toxic,* are however, rarely desirable in any product stream, thus their in-situ use usually necessitates that they be completely removed prior to product use, which is generally difficult in high linear-velocity processing: Because these particles are so small (and numerous), very robust (e.g. N.F or R.O.) operations would often be required to assure such removal, rendering most in-situ HTCT3 nanocatalytic operations extremely cost-prohibitive. Conversely, ex-situ use of nanocatalysts (immobilizing them in such a way as to thoroughly exploit their advantages in a flow-through reactor both safely and cost effectively) has been stymied as well. Why? Until the advent of High-Efficiency Nano-Catalyst Immobilization (H.E.N.C.I.) technology, the following seven engineering criteria inherent to immobilizing nano-sized particles for cost-effective ex-situ HTCT3 catalysis had proven insurmountable:

1. **Complete Immobilization – nano-catalytic particles are held 'fast' in place and never become part of the reactant matrix phase as they effect reaction(s):** The particles never actually enter into the reactant solution, and thus never have to be removed via subsequent expensive and elaborate unit operations such as R.O., NF.& mag. filtering. The process can be validated to emit zero nano- /micro- particles in the effluent.

2. **Ultra High Immobilization Density:** At ~10^{10} to 10^{13} particles per gram, many nanocatalysts must indeed be immobilized per unit reactor volume in order for the technology to be truly valuable for HTCT3 applications. In preliminary tests using ~80mn Zero-Valent Iron Fe/Pd nanocatalyst particles HENCI reactors easily be immobilized over 25 x 10^{15} particles - 3.7 million ft^2 of available catalytic surface area per cubic foot of reactor volume, without significant pressure drop through the reactor, at linear velocities in excess of 60 FPM.
3. **High Mass Transport Efficiency:** When micro-or nano-sized structures are employed for catalysis, boundary layers are virtually eliminated, and nearly 100% of the surface area of each catalytic particle is directly exposed to the bulk-composition reactant matrix (low-transport-rate pore-wall volume is negligible or non-existent on nano-structures – all surface area is on the particle perimeter). Moreover, HENCI technology overcomes the natural tendency for small-particle agglomeration so that the particles are immobilized within the HENCI reactor in *a micro-homogeneous three-dimensional mono-disperse array* of individual particles within the column, eliminating flow-channeling and maximizing overall reaction rates. The static micro-mixing effect of HENCI facilitates overall rates equal to or greater than those achievable via in-situ agitation, and 'plug' flow often occurs at superficial *Nre* under 300.
4. **High Momentum Transport Efficiency:** HENCI columns exhibit extremely low pressure-drop per unit reactant matrix flux: linear velocities of hundreds of ft/second can be achieved with extremely small pressure drops per unit reactor length, and hence with small, inexpensive pumps. *For GWR, the head available at the source is usually more than adequate, eliminating the need for additional pumps altogether*
5. **The Technology is Readily Scalable & Configurable:** Any combination of throughput & conversion can be addressed with HENCI. System modularity is easily accomplished, minimizing system capital costs.
6. **The Technology is Inexpensive to Build and Operate:** On the basis of mass of nanocatalyst immobilized, HENCI facilitated systems are roughly 1 to 2 orders of magnitude less expensive to manufacture, operate, and maintain than any other immobilization technology known by the author.
7. **Potential for quick regulatory-agency acceptance:** HENCI usually exposes the reactant matrix to no additional contact materials (beyond those in the catalyst particles themselves).

HENCI technology alone meets these criteria, thus comprising the 'last piece in the puzzle' of unleashing the true power of nanocatalysis outside the lab by safely exploiting it's highly favorable kinetic and transport rates, especially in HTCT3 applications such as large-scale remediation of our now-carcinogenic groundwater supplies (see below). The micro-homogeneous, ultra-high-density, ultra-low pressure-drop dispersion accomplished within the tubular HENCI reactor has come to be known also as the Dispersed Nano/Micro-Particle Reactor Bed, or **HENCI DNP reactor** for short. Dr. Mamadou Diallo at the Beckman Institute/Power, Environmental & Energy Research Center, both of Cal Tech; Dr. Michael Wong et. al. at CBEN / CNST, Rice University, and Drs. Peter Rony, Dr. John Y Walz Jr., and Dr. Preston Durrill et. al. of Virginia Tech's Ch.E. department are currently evaluating and developing new GWR and CPI applications and process scenarios using HENCI DNP units. Several other universities are also in the process of obtaining HENCI DNP systems.

One application deserves note: Groundwater Remediation. Though CHC levels in many of the thousands of polluted ground water basins may be but a few ppm, many of these ubiquitous carcinogenics have EPA limits under 15 ppb for potability, and hence often require 99.9% (or higher) conversion to achieve regulatory compliance. Thus, heretofore, large-scale (High throughput x High conversion)catalytic groundwater remediation (GWR) has remained infeasible due to the mass-transport limitations inherent to trace pollutants competing for reaction-sites when 'outnumbered' by the H2O matrix (for every billion catalyst-surface sorption events only a few actually involve the reactant). However, today, when the right nano-particle catalyst composition[1] is charged to an HENCI DNP system, rapid and complete catalytic destruction (cleavage/breakdown to benign species) of dozens of CHC and other toxins, including the perchlorate anion [2] can be effected with unprecedented process efficiencies: Per unit throughput of remediated water, catalytic groundwater remediation plant sizes, operating, and capital cost are reduced drastically, and thus brought, for the first time ever, well into the realm of feasibility

Bibliography

1: Zhang, W.; ***Journal of Nanoparticle*** *Research* 5: 323–332, 2003.

2: Pirkanniemi; ***Chemosphere 48 (2002) 1047–1060***

3: Jiasheng Cao, et.al.; Journal of Nanoparticle Research (2005) 7: 499–506

4: Schrick, Blough et. al., *Chem. Mater.* 2002, *14,* 5140-5147

IRON/PALLADIUM NANOPARTICLE – CATALYZED BREAKDOWN OF TRICHLORO-ETHYLENE IN DILUTE AQUEOUS SOLUTION IN A CONTINUOUS H.E.N.C.I. REACTOR

Kenneth Cross, Ch.E.
Cross Technologies, Jan 2005

The Efficacy of Fe/Pd nanoparticles for breakdown of commonly found recalcitrant halogenated hydrocarbon groundwater pollutants is now widely appreciated, yet no commercially viable ex-situ process for such breakdown exists due to the difficulties inherent in 'fixing' or immobilizing these nano-scale catalyst particles in a reactor bed: All existing nano-catalyst immobilization (NCI) techniques are either themselves cost-prohibitive (on a per-pound-nanocatalyst basis) or result in thermodynamic and kinetic reaction efficiencies so low as to render the resulting process model grossly uneconomical for any but the most exotic applications, and certainly for 'commodity' type operations such as water remediation. The following experiment was designed to quickly test the kinetic efficiency and cost effectiveness of the new High-Efficiency Nano-Catalyst Immobilization (HENCI) technology by Cross Consulting & Engineering. GAC-filtered tap water artificially – contaminated with trichloroethylene (TCE) was pumped through pilot-scale HENCI reactors containing specially-immobilized Fe/Pd nanoparticles (nanocatalysts) at ambient pressure / temperature in order to determine the efficacy (e.g. the overall conversion-reactor efficiency) of these Fe/Pd nanoparticles in catalyzing the breakdown TCE when immobilized via the HENCI method.

In order to determine the amount of reduction in TCE concentration due to the process, samples of the artificially polluted water were taken before treatment and after each of three sequential 'passes' through the HENCI reactor, and then submitted for quantitative analysis of the pollutants studied. Though these preliminary experiments were conducted in a manner which minimized evaporation of the TCE from the sample water in the pilot system during the runs, financial constraints precluded running the tests such that the pollutant evaporation during the tests was completely eliminated, and hence evaporation of TCE is therefore responsible for some fraction of the decrease in concentration observed throughout the runs. During these very short catalyst-exposure times (~1.5 seconds per pass) however, the sum of (1) the nanoparticle-catalyzed breakdown of trichloroethylene and (2) the evaporation of TCE was approximately 30% per 1.5-second pass through ~150 g of the HENCI-immobilized nanocatalyst at an initial TCE concentration of ~140 ppm.

TABLE OF CONTENTS

Introduction And Theory

During the last 50 years, dozens if not hundreds of aquifers across the continental US have been contaminated with various halogenated and oxy-halogenated hydrocarbons, largely through military and industrial operations. Hence, a significant amount of work has been done on the development of technologies for the remediation of such groundwater. The efficacy of bimetallic catalysis for dehalogenation and reduction of such compounds is well established. The mechanism effecting catalysis (lowering of the reaction E_A) is a subject of considerable study. Hydrogen- and electron-transport halide-reduction mechanisms have been suggested, both of which seem to occur as the ZVI is an effective electron donor, and the H+ ions are transported by the more noble meals. In addition, the resulting potential across the interface of the two metal species arising from the difference in the standard electrode potentials of the two metals may well play a part, as may other forces which tend to strain the carbon-halogen covalent bonds. In the case of Iron / Palladium catalyst particles, the standard electrode potential (hereafter SHE) of palladium is +0.987 V SHE and that of Iron (Fe) is –0.440V SHE. This means that when the two metals are brought into intimate contact, the resulting bimetallic voltage at their interface is approximately Del V= 1.427 V. This voltage, along with other small forces which a molecule incurs at the bimetallic and Palladium – water interface, has a catalytic action on the species in question.

In any event, the combined effect of these forces is to break the carbon-halogen bond(s) in question, cleaving the molecule and releasing the halogen as a radical or ion and causing the olefin to hydrogenate or hydroxylate at that site or form a higher-level carbon - carbon bond with a neighboring vinyl group. The halogen radicals and ions then tend to readily combine with one another, hydrogen ions, or metallic species in the water to form diatomic halogen species, acids, and salts respectively, all of which have very low specific Gibbs free energies and are thus extremely stable. **In general, the resulting species are more environmentally benign and remediation is effected.** In order for the reaction to take place, however, the hydrocarbon pollutant species must come into intimate contact with the catalyst particles' surfaces, and possibly even the actual bimetallic interface. When the pollutant exists in an aqueous solution at the ppm or ppb level, however, the transport of the molecule to the bimetallic interface of a standard-sized catalytic particle becomes (by far) the rate-limiting step, and previous attempts to remediate polluted water have not been practicable because the product "P" of

P = (AREA OF BIMETALLIC INTERFACE AVAILABLE x REACTOR RESIDENCE TIME x INLET POLLUTANT CONCENTRATION)

has been far to small , and most of the pollutant molecules passing through a typical packed reactor never actually adsorb onto the interface that they may be broken down. The only way to achieve satisfactory overall conversion is to greatly increase the product "P". Since decreasing the pollutant concentration is the object of remediation, increasing P boils down to increasing the bimetallic area by making and immobilizing smaller bimetallic catalyst particles and / or designing reactors with higher residence times. So when the former was accomplished by Golder and associates, the authors set out to find a way to immobilize and expose a very large number of these bimetallic nanoparticles (hereafter BNP's) in a small volume, that each was available to contact over nearly all of it's surface area AND fixed in place such that it would not become entrained in the flow through the reactor. Once a suitable method of immobilization and fixation was developed and tested, the feasibility of this 'pump-treat-and-use' scheme was greatly enhanced, and the preliminary tests herein were warranted. After the immobilization method was demonstrated, the experimental protocol was developed. Since the method of immobilization was not conducive to accurate modeling, an empirical approach to the tests was adopted: A fully-blocked (non-orthogonal) matrix of test runs was first proposed to explore the efficacy of each of two reactors in remediating small batches of artificially polluted water under varying conditions. The parameters varied included: two values of flowrate to study the efficacy of the system at different residence times and two different inlet contaminant concentrations. Varying these data would have varied the above-mentioned product "P" to provide insight into the true design constraints determining the economic feasibility of differing strategies for site remediation via this pump-and-treat method. The protocol, however, called for some 96 runs to be made for each reactor, totaling 192 runs and requiring 240 separate sample analyses to complete the study. At an approximate average cost of ~$125 per analysis, the projected cost of obtaining response data from such a test (~$30,000.) was deemed prohibitive, and a much less comprehensive experiment was instead carried out.

Apparatus

Referring to figure 1, the apparatus included a covered feed container into which a submersible pump was lowered. The pump pumped the fluid from the feed container through the reactor, through a variable area flow meter, and into the open discharge container. The discharge and feed containers were interchangeable to facilitate additional 'passes' through the reactor (2nd and 3rd-stage treatments) without exposing the water to the atmosphere unnecessarily, effectively allowing us to study, once again, different values of our product "P".

Figures

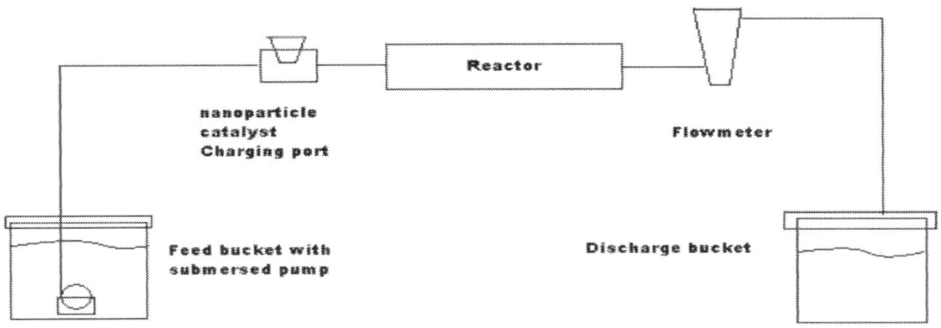

Procedure

1) Protocol

a) P1 reactor was challenged with two pollutants. Then the P2 reactor was challenged with three pollutants, making 5 runs in all, as depicted in the following experiment Matrix:

2) Contaminant Species	3) Reactor Type	4) Pass #	5) Flowrate, GPM
6) Trichloroethylene	7) P1	8) 1	9) 1
10)	11)	12) 2	13) 1
14)	15)	16) 3	17) 1
18) Ammonium Perchlorate	19) P1	20) 1	21) 1
22)	23)	24) 2	25) 1
26)	27)	28) 3	29) 1
30) Trichloroethylene	31) P2	32) 1	33) 0.28
34)	35)	36) 2	37) 0.26
38)	39)	40) 3	41) 0.25
42) Ammonium Perchlorate	43) P2	44) 1	45) 0.28
46)	47)	48) 2	49) 0.25
50)	51)	52) 3	53) 0.24
54) N-Nitrosodimethylamine	55) P2	56) 1	57) 0.22
58)	59)	60) 2	61) 0.25
62)	63)	64) 3	65) 0.25

66) Method

b) The aqueous BNP solution was diluted by 50 % due to H2O evaporation which had taken place after purchase and before testing: 250 ml filtered H2O was added to the 250 ml NP solution remaining in the original bottle. The bottle was then vigorously shaken to homogenize the BNP's

c) The P1 reactor was prepared for service and connected into reaction system
   i) The HENCI reactor was prepared
   ii) The reactor was sealed off by closing it's ISO valves
   iii) 3l water added to sealed reactor
   iv) 100 ml NP solution was added to sealed reactor

d) The recycle tube was then plumbed in (bypassing flow meter , and leading back into the dummy feed bucket filled with flush-water)

e) ISO valves were opened

f) Pump was turned on and flush water recycled through reactor for 5 minutes to allow all NPs to become bound in media

g) Pump turned off and ISO valves closed

h) Flush water allowed to drain from system. About 1 l flush water remains in reactor and tubing.

i) Switched feed bucket to that with the pollutant solution
j) Turned on pump and recycled pollutant solution directly back into feed bucket for 30 to 60 sec
k) Took sample of the feed bucket solution for analysis
l) Conduct First Pass test
    i) Close recycle valve and open ISO valves to begin flow through the rector
    ii) Adjust flow meter valve to achieve desired flowrate
    iii) Allow reactor discharge to collect win discharge bucket until feed bucket is empty
    iv) Sample from discharge bucket and label sample appropriately (for whichever pass)
    v) Discard feed bucket
    vi) Carefully bring discharge bucket over to feed-side of the apparatus, insert the submersible pump, and close the lid. This bucket is now the feed bucket
    vii) Install a new bucket under the discharge tube of the process. This bucket is now the discharge bucket
m) Conduct Second Pass
    i) Repeat steps k)I through k) VII above
n) Conduct Third Pass
    i) Repeat steps k)I through k) VII above
o) Clean all equipment and repeat steps c) through m) above for the next pollutant (see matrix)
p) The P2 reactor was prepared for service and connected into reaction system
    i) The HENCI reactor was prepared
    ii) The reactor was sealed off by closing it's ISO valves
    iii) ~1l water added to sealed reactor
    iv) Reactor BNP retention system was enabled
    v) 30 ml BNP solution was added to sealed reactor
q) Steps c) through n) were repeated for the P2 reactor for it's first pollutant challenge (see test matrix)
r) Steps c) through n) were repeated for the P2 reactor for it's second pollutant challenge (see test matrix)
s) Steps c) through n) were repeated for the P2 reactor for it's third pollutant challenge (see test matrix)
t) Equipment was cleaned, disassembled and packed up

Results

The analyses of the 20 samples along with a few calculated values are summarized below. Unfortunately, for reasons beyond the scope of this paper, **only the third quartet of data (TCE on the P2 reactor – red and *italicized*) are valid response data**. Hence, all discussion of the results will be limited to this sole valid dataset.

Input Variables and Raw Data							Calculated Values	
Contaminant Species	Reactor Type	Flowrate, GPM	Calc'd Inlet Conc, mg/l	Measured Inlet Conc, mg/l	Pass #	Measured DischargeConc, mg/l	Reaction Efficiency (% conversion)	Overall (3-pass) Percent Conversion
Trichloroethylene	P1	1	148.000	8.240	1st	15.300	N/A	
		1			2nd	17.400	N/A	
		1			3rd	12.300	N/A	N/A
Ammonium Perchlorate	P1	1	~120	130.000	1st	81.000	37.69	
		1			2nd	100.000	-23.46	
		1			3rd	92.000	8.00	29.23
Trichloroethylene	***P2***	***0.28***	***148.000***	***14.200***	***1st***	***10.300***	***27.46***	
		0.26			***2nd***	***6.970***	***32.33***	
		0.25			***3rd***	***5.080***	***27.12***	***64.23***

Ammonium Perchlorate P2	0.28	~100	94.000	1st		88.000	6.38	
	0.25			2nd		88.000	0.00	
	0.24			3rd		89.000	-1.14	5.32
N-Nitrosodimethylamine P2	0.22	100.004	ND	1st	ND			
	0.25			2nd	ND			
	0.25			3rd	ND			

Discussion

1) Data Trend Analysis

a) **TCE on the P1 Reactor** – no valid response data
b) **Ammonium Perchlorate on the P1 Reactor** – no valid response data
c) **TCE on the P2 reactor**:
  i) **Overview:** Evidently, in the TCE run on the P1 reactor, each pass through the reactor decreased the concentration of Trichloroethylene by approximately 30% (see data table). This means that the combined effect of evaporation and catalytic breakdown of TCE during each test run was to decrease the concentration of TCE by ~30 % per pass.
  ii) **Reaction and Mechanism:** Since the reaction of interest is the breakdown of TCE, we shall take a look at the reaction kinetics and see if this data can corroborate a valid kinetic model. First let's take a look at the reaction and it's mechanism(s).
  (1) For the purposes of this report, we'll assume that the pH of the water is between 5 and 9 so that

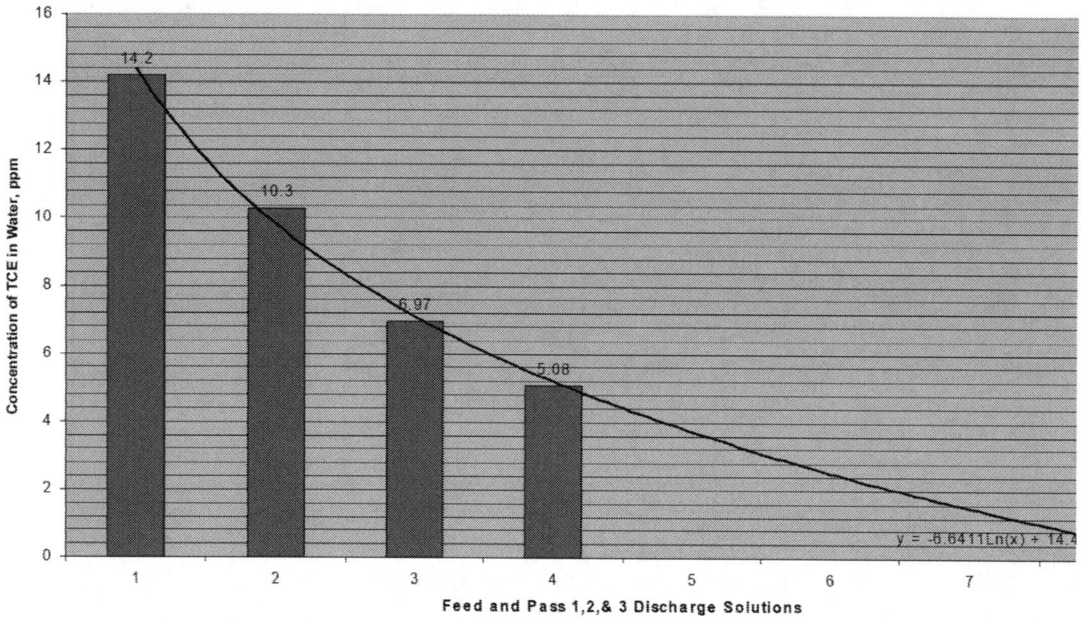

no Lewis-acid/base reactions occur with a high enough rate to compete with the cleavage reaction itself. It is the author's understanding that this assumption is most often valid within the groundwater remediation industry.
  (2) In this regime, when TCE comes into contact with the bimetallic interface and chlorine is removed from the aliphatic, it may well receive an extra electron from the (cathodic) palladium or zero-valent iron during, or to facilitate, the cleavage reaction. In that case, the resulting chlorine anion (Cl^-) may be more likely to take place in an ionic combination: It may recombine

with (1) a metal cation in solution to create a salt (a fairly benign fate for Cl in most treatment applications), or (2), a hydrogen cation to create the Lewis acid HCl, and the overall effect of the extra electrons from the palladium may be to lower the pH of the effluent, if it is indeed high at the reactor inlet (another positive outcome or 'fate' for the Cl). If it does not, however, receive an extra electron upon cleavage and it 'comes off' as a radical ($Cl^{\cdot}$), then it is more likely to recombine into Cl2 gas. In either case, however, note that the olefin itself has not had to supply the extra electron, and therefore is likely to remain an alkene: that is , the double bond will likely not be broken to a single (alkane) bond. However, if the Palladium does NOT act as a true cathode, then it is quite possible that the double bond will be reduced to a single C-C bond and an alkane (ethane) can be formed. This means that there are now six free bond-sites to be filled on each ethane molecule created. If this DOES happen, and if the pH of the solution is high enough, it is quite conceivable that one or more of the free sites will be replaced with hydroxyl ions instead of hydrogen ions, resulting in an alcohol or poly-ol, some of which, like ethanol, may be more benign than ethane. Other (higher-order) polyols, however, may be less benign than ethane.

(3) In either case, the bond-order of the resulting aliphatic hydrocarbon is much less toxicologically important than the degree of dehalogenation: *it is the conversion to a completely dehalogenated species which is most important*: the partial dehalogenation of TCE to dichloro- or the (less likely) chloro-ethene (hereafter DCE and MCE) is of little good, as these compounds are also carcinogenic and poisonous. Hence, one of the biggest weaknesses of this study is that we do not assay for DCE or MCE: hence from our results, we *can only ascertain how much TCE is gone – not what we have replaced it with*. In other words, we can only calculate percent breakdown, not percent conversion.

(4) With a bit of up-stream (automated) pretreatment and more research, however, the potential to convert TCE very *completely* to very benign breakdown byproducts and to verify the conversion is very real

(5) In any case, for our purposes here, let's look at the portion of the cleavage reactions which DO go to completion, and are re-substituted by only hydrogen , resulting in the completely unsaturated byproduct ethylene, leaving study of conversion optimization for a future round of tests.

   (a) The overall cleavage reaction can be written thusly: $2(C_2Cl_3H) + 6H^+ \rightarrow 2(C_2H_4) + 3Cl_2$.

   (b) Now, the reaction mechanism may well look something like this:

(6) $C_2Cl_3H \rightarrow C2H + 3Cl$

(7) $C2H + 3H+ \rightarrow C2H4$

(8) And therein lies the rub: which of the two is the slowest (rate-limiting) step? If step (1) is rate-limiting, then the reaction kinetics can be estimated as first-order, meaning that the rate of conversion of TCE into ethylene is simply proportional to the concentration of TCE in the bulk of the reactor, and hence, at steady-state operation, proportional to the concentration of TCE in the feed stream. If however, step (2) is rate-limiting, then the conversion rate will be proportional to the inverse log of the pH of the feed stream. Since we've no control over the initial feed stream of polluted groundwater, and furthermore, whereas we could adjust the pH of our reactor inlet stream quite easily with acid or base injections, then for the purposes of this discussion, let us explore what kind of data correlation we get assuming that pH of the inlet stream is constant at some value which provides enough hydrogen ions in solution so as to make step 1 above the rate-limiting step

iii) ***Kinetics:*** In this case, then, we have the dehalogenation itself as the rate-limiting step. Now, since the cis-and trans- Cls have differing bond-energies arising from the stereochemistry of the molecule, they will not all be cleaved with the same 'ease'. Hence, the actual mechanism may well include

DCE and MCE intermediaries, as the TCE loses one, then two, then three of it's chlorine 'ligands' sequentially. Assuming, however, that the cleavage reaction does go to completion at each reaction site (the interface of Pd and Fe on the nanoparticle surface) and all three chlorine ligands are removed before the hydrocarbon leaves (desorbs or ' vibrates off') the BNP surface and back into solution, then we can develop a model for the (first-order) kinetics of the reaction: The rate equations can be written thusly:

iv) $-d[TCE]/dt = k \times [TCE]$ which , upon rearrangement yields

v) $d[TCE]/[TCE] = -kdt$ which upon integration yields

vi) $\ln[TCE] - \ln[TCE]_0 = -k(t-t_0) = -kt$ which, upon rearrangement yields

vii) $\ln[TCE] = \ln[TCE]_0 - kt$ which, upon taking the inverse natural log of each side, becomes : $\mathbf{[TCE] = [TCE]_0 \times e^{-kt}}$

viii) **If we then fit these four data points into this math model, $k = 95.5\ hr^{-1}$. (see accompanying excel spreadsheet where this data is correlated), which is about twice that which occurs when the catalyst is used in-situ. In other words, the use of a HENSI Nano-disperse bed actually has a much greater mixing effect than does rolling or shaking a batch of reactant with in-situ catalyst particles.** Note that this is the typical Ahhrenius form of first order decay kinetics: The *rate* of conversion decreases as the amount of TCE *available* for conversion decreases. Now, if we try substituting values of, say 18 ppm for $[TCE]_0$, and 0.3 for k, and then graph this equation, we get the familiar first-order 'Ahhrenius' curve with the following values:

1st Order Rate Curve [TCE] = [TCE]0 x eEXP -0.3t

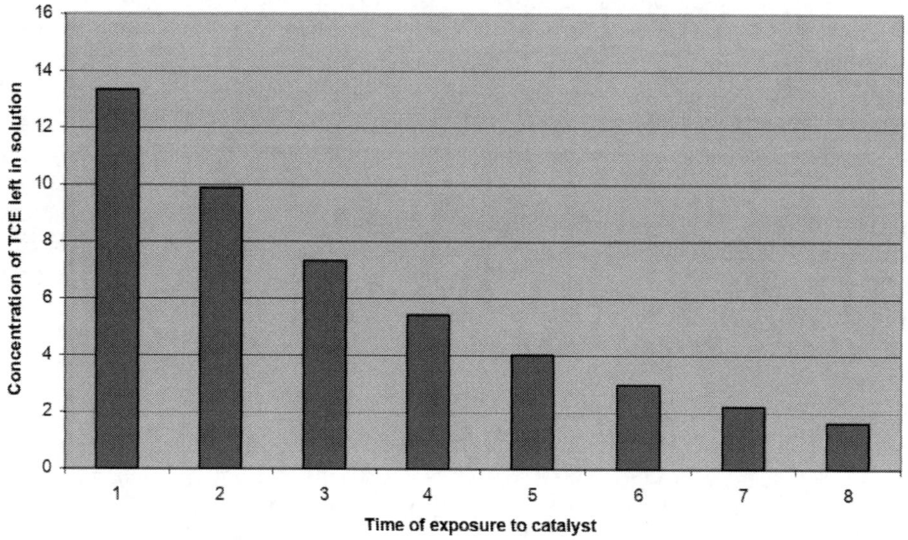

ix) Note how similar in shape (and in values!) this curve looks to the data taken at the MNWD for the P2 reactor breaking down TCE. The main difference is that this model would be best applied to a batch reactor wherein the TCE solution was simply exposed to the catalyst for an extended period of time (called 'batch exposure time') *without flowing*, and then the perfect 'average' sample was magically taken out of the solution every so often. This situation is effectively mimicked, however, by our flow-through scheme (called a ***continuous*** reactor as opposed to a *batch* reactor) in which the samples were taken after each 'run'.

x) The data are summarized in the following xls file:

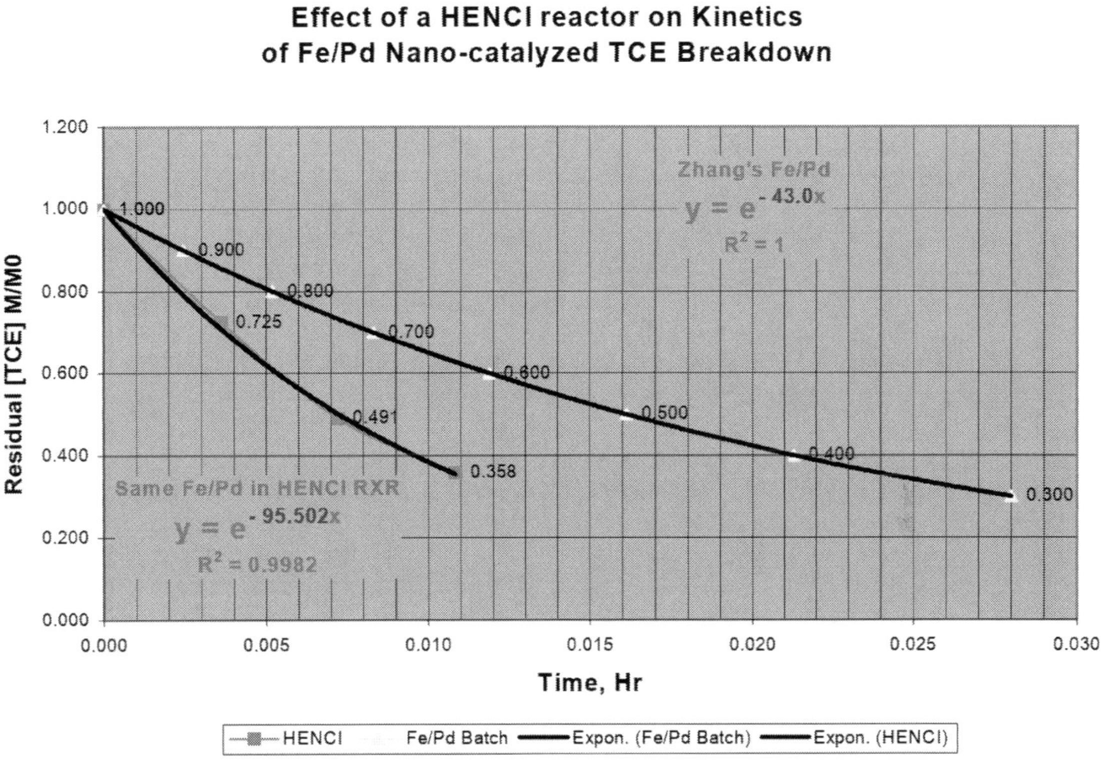

xi) ***Process Engineering Analysis and Scale-up:*** In fact, it is important to understand the continuous-reactor's counterpart to 'batch exposure time': the duration of exposure for each molecule of water or TCE as it flowed through the reactor. This is called the reactor 'Residence Time', and is only completely valid number when the reactor is operated at fully developed (turbulent or 'plug') flow , which only occurs when the reactor's Reynolds number (ρvd/u) is over 2300. In our reactor, however, we can relax this requirement because physical conditions in the reactor render the flow effectively turbulent at low superficial values of the Reynolds number, though without increasing the reactor's overall pressure drop enough to impact commercial applicability. It is important to point out that, even with this effect, it is likely that the distribution through our reactor is not perfectly 'flat': some molecules may have spent significantly longer in the reactor than others. This problem is easy to remedy, however, increasing the length of the reactor and the flow rate simultaneously. These changes are important for the next round of tests and will be carefully made because once we can confidently assume plug flow through our reactor(s), our analyses will be easier AND more valid. Again, only further testing can determine the optimal values of flowrate and reactor length, not to mention a host of other parameters beyond the scope here. For our purposes here then, we'll use the median value of residence time to represent the true residence time of the reactor. The residence time is then the volume of the reactor divided by the flowrate:

xii) **Residence Time (Tr) = Reactor Volume / Flowrate**. Since the valid data set was obtained on the P2 reactor, we'll calculate the Tr for the P2 reactor. The P2 reactor volume is 15.086 in^3 = 0.0653074 US gal, but our best estimate of the portion of the reactor volume occupied by the packing and NPs is about 18%, so the actual reactor volume is 0.82x .00653 = 0.0535521 gal. Now, from the data table above, we know that the flowrate through the P2 reactor averaged approximately

0.26 GPM, so Tr = 0.206 min, or about 12.4 seconds. **So we achieved approximately 30% removal of TCE (evaporation as well as reaction) in about 12 seconds of total exposure at an inlet concentration of about 14 ppm.** Furthermore, the percent conversion at the lower inlet concentrations in the data (the last run achieves ~30% removal at about 5 ppm / 14 ppm or 36% of the inlet concentration) stays about the same, suggesting that the conversion of TCE to ethane and Cl2 gas is not mass-transport-limited at these inlet concentrations. If this also holds true at concentrations of about 1/3rd of what we now see, we'll know that we can take 1.4 ppm down to less than 14 ppb in 13 reactor lengths or about 20 ft of tubing, a length which can easily be wrapped into a column small enough to fit on the back of a small truck. Of course, scaling up the nanoparticle retention per unit reactor volume *and* the diameter of the reactor are also feasible and would result in significant increases in conversion. For instance, if we assume that doubling the nanoparticle loading (surface area available for reaction / vol. of the reactor) would increase the rate of destruction by even 80% (a safe bet), and we also increase the diameter of the reactor to, say, 1 ft, then , nominally, due to the increased conversion and throughput, it turns out that for this particular scale-up, **an 8-ft long , 1 ft diameter reactor would be able to remediate (to < 1.0 ppb) approximately 30 GPM of 1.4 ppm TCE inlet water.** This now represents a sensible and economically feasible pump-and-treat solution which operates at the pollutant levels found in groundwater and can be mobilized on a small truck.

xiii) ***Other Considerations:*** These results seem commensurate with previous research on the in-situ remediation of TCE with Fe / Pd nanoparticles (see Zhang, Wei-xian, Journal of Nanoparticle Research, 5:323-332, 2003) which attests to the ability of these nanoparticles to remediate TCE (and similarly) polluted groundwater to <10 ppb TCE within 8 hrs of batch exposure at a nanoparticle loading concentration of 6.25g BNP's per liter of polluted water. The extended time period is necessary to allow each TCE molecule in the quiescent body of water (whether it be an aquifer or a beaker) to come into contact (through random Brownian motion) with the nanoparticle bimetallic interface. The configuration of the continuous reactors used herein provide a 'boost' to that Brownian- (and concentration gradient-) driven contact by greatly increasing the coefficient of mass transport for the reaction system in the same way that a packed column increases overall mass transport over a non-stirred batch reactor.

d) Ammonium Perchlorate on the P2 Reactor – no valid response data

e) Nitrosodimethylamine on the P2 Reactor – no valid response data.

Conclusions

i) Though it is becoming widely known that Bi- and even mono-metallic zero-valent Fe-containing nanoparticles hold much promise for groundwater remediation, with the new method of immobilization used herein, new life may be afforded to pump-and treat strategies for groundwater remediation.

ii) Specialized bimetallic nanoparticles also have a great potential in general catalysis in the chemical process industries. The proprietary nanoparticle immobilization techniques used in this study will be of great value to an assortment of high-tech and Ultra-High-Purity chemical process-based industries.

iii) In order for a continuous reactor approach to the exploitation of BNP catalytic efficacy to be truly advantageous, the continuous reactor configuration must be correctly designed to affect intimate contact of the fluid to the BNP's with each pass through the reactor body. In this way only, the continuous reactor has the effect of greatly increasing the effective loading of BNP's per unit volume of water without actually using up the BNP's, and the overall rate of chemical conversion is greatly enhanced

iv)

References

Zhang, Wei-xian, Journal of Nanoparticle Research, 5:323-332, 2003

Ammonia by Pressure Swing Adsorption

Felix Jegede, SmartKoncept Technology, 11999 Katy Frwy,, Houston,, TX 77079
and
Kent S. Knaebel, Adsorption Research Inc., 6175-D Shamrock Court, Dublin, OH 43016.

Abstract
Ammonia is produced by passing a heated compressed mixture of Nitrogen and Hydrogen over a catalyst, at a pressure of between 80 – 220 bar, and a temperature of between 450 - 550°C. Recovery of ammonia is traditionally achieved by refrigeration, i.e., by cooling the effluent gas mixture to a relatively low temperature of about -5°C to about -40°C. Refrigeration is energy intensive and does not lead to complete recovery of the ammonia. The un-recovered ammonia along with unconverted reactants and inerts are recompressed and recycled back to the reactor. This further reduces the reactor efficiency or yield. In this project, we propose to develop and demonstrate a technically feasible and commercially viable system that integrates reaction to produce ammonia along with recovery of the products by adsorption separation methods and significantly decrease the energy requirement in ammonia production. The proposed method can be applied to both green plants process as well as to retrofit and revamp existing ammonia plants in the U.S and achieve significant reduction in the energy requirement, raw material usage, capital cost, and environmental burdens while enhancing safety of industrial ammonia production. As part of this project, we will also examine the potential effects of high temperature and pressure on adsorbent stability in an Ammonia Process.

Introduction

Ammonia, (NH_3), is widely used in the industry for a variety of purposes including Power plants (at least 225 of the 275 electric utilities and power plants in the USA); Fertilizer production (including ammonium sulfate, ammonium phosphate, ammonium nitrate); Chemicals synthesis (such as nitric acid, sodium hydrogen carbonate, hydrogen cyanide, as well as hydrazine), Fibers & Plastics (such nylon), Pharmaceutical drugs (such as sulfonamide), Mining&Metallurgy, and in industrial and household cleaning agents. The worldwide production in 2004 was 109,000,000 metric tons.[1] China produced 28.4% of the worldwide production followed by India with 8.6%, Russia with 8.4%, and the United States with 8.2%.

Recent DOE-EERE-ITP Chemicals Subprogram studies have consistently ranked ethylene and ammonia as the top two most energy intensive processes. Ammonia production consumes a total energy of 750 Trillion BTU, (representing more than 10% of total energy in production of top 50 major chemicals of 7,220 Tril Btu). Many of ammonia derivatives are also ranked among the top 50 (including nitric acid, ammonium nitrate, ammonium sulphate and urea).

New innovative methods are needed to address and improve (the chemistry and separation methods of) older processes that are more energy efficient. This project focuses on development of a more energy efficient way to produce ammonia and aligns well with the U.S. chemical industry's need.

Objective of this Project

In this project, we propose to develop and demonstrate a technically feasible and commercially viable system that integrates reaction to produce ammonia along with recovery of the products by adsorption separation methods and significantly decrease the energy requirement in ammonia production. The proposed method can be applied to both green plants process as well as to retrofit and revamp existing ammonia plants in the U.S and achieve significant reduction in the energy requirement, raw material usage, capital cost, and environmental burdens while enhancing safety of industrial ammonia production. As part of this project, we will also examine the potential effects of high temperature and pressure on adsorbent stability in an Ammonia Process.

Ammonia Synthesis.
The conditions under which Nitrogen and hydrogen would combine to produce ammonia were first established in 1909 by Fritz Haber. .[2,3,4] These conditions include, medium temperature (~500°C) ; very high pressure (~250 atmospheres, ~351kPa) ; a catalyst (a porous iron catalyst prepared by reducing magnetite, Fe_3O_4). The Haber synthesis was subsequently developed into an industrial process by Carl Bosch.

A typical modern ammonia-producing plant first converts natural gas (i.e., methane) or LPG (liquified petroleum gases such as propane and butane) or petroleum naphtha into gaseous hydrogen. The method for producing hydrogen from hydrocarbons is referred to as "Steam Reforming".[5] The purified Nitrogen and Hydrogen are mixed in the right proportion to form a mixture of gaseous nitrogen and hydrogen in a 1 to 3 molar[6].

Starting with a natural gas feedstock, the processes used in producing the hydrogen are:

(i) to remove sulfur compounds from the feedstock because sulfur deactivates the catalysts used in subsequent steps. Sulfur removal requires catalytic hydrogenation to convert sulfur compounds in the feedstocks to gaseous hydrogen sulfide:
$H2 + RSH \rightarrow RH + H2S(gas)$

(ii) The gaseous hydrogen sulfide is then absorbed and removed by passing it through beds of zinc oxide where it is converted to solid zinc sulfide:
$H2S + ZnO \rightarrow ZnS + H2O$

(iii) Catalytic steam reforming of the sulfur-free feedstock is then used to form hydrogen plus carbon monoxide:
$CH4 + H2O \rightarrow CO + 3H2$

(iv) The next step then uses catalytic shift conversion to convert the carbon monoxide to carbon dioxide and more hydrogen:
$CO + H2O \rightarrow CO2 + H2$

(v) The carbon dioxide is then removed either by absorption in aqueous ethanolamine solutions or by adsorption in pressure swing adsorbers (PSA) using proprietary solid adsorption media.

(vi) The final step in producing the hydrogen is to use catalytic methanation to remove any small residual amounts of carbon monoxide or carbon dioxide from the hydrogen:
$CO + 3H2 \rightarrow CH4 + H2O$
$CO2 + 4H2 \rightarrow CH4 + 2H2O$

(vii) To produce the desired end-product ammonia, the hydrogen is then catalytically reacted with nitrogen (derived from process air) to form anhydrous liquid ammonia. This step is known as the ammonia synthesis loop (also referred to as the Haber-Bosch process):

The steam reforming, shift conversion, carbon dioxide removal and methanation steps each operate at absolute pressures of about 25 to 35 bar, and the ammonia synthesis loop operates at absolute pressures ranging from 80 to 220 bar depending upon which proprietary design is used.

Typically, this gas mixture (of Nitrogen & Hydrogen) is compressed to high pressure. The decision on the pressure at which to carry out the ammonia synthesis is an economic factor, generally this pressure being in a range from 150 to 1000 atmospheres (e.g., about 80 bar to about 220 bar). The compressed Nitrogen and Hydrogen mixture is heated at a temperature of about 450 to 550 °C and passed over a catalyst. Some ammonia formation occurs.

$$N2 + 3H2 = 2NH3 \qquad H = -92.4 \text{ kJ mol}^{-1}$$
$$H = H_{(products)} - H_{(reactants)} = \text{negative (-ve)}$$

The reaction between nitrogen gas and hydrogen gas to produce ammonia gas is exothermic, releasing 92.4kJ/mol of energy at 298K (25°C).

The reaction is an equilibrium reaction, with a rate constant of:

$$K_{eq} = \frac{[NH_3]^2}{[N_2][H_2]^3}$$

where rate constants are:

Temp. (°C)	Keq
25	6.4×10^{2}
200	4.4×10^{-1}
300	4.3×10^{-3}
400	1.6×10^{-4}
500	1.5×10^{-5}

The industrial yield from the reactor is typically 10-20% yield depending on what catalyst is used. The proportion of ammonia formed depends on the temperature and pressure used, the activity of the catalyst and time of contact of the gases with the catalyst. By Le Chetalier's Principle:

- decreasing the temperature causes the equilibrium position to move to the right resulting in a higher yield of ammonia since the reaction is exothermic (releases heat). Reducing the temperature means the system will adjust to minimize the effect of the change, that is, it will produce more heat since energy is a product of the reaction, and will therefore produce more ammonia gas as well. However, the rate of the reaction at lower temperatures is extremely slow, so a higher temperature must be used to speed up the reaction which results in a lower yield of ammonia.
- increasing the pressure causes the equilibrium position to move to the right resulting in a higher yield of ammonia since there are more gas molecules on the left hand side of the equation (4 in total) than there are on the right hand side of the equation. Increasing the pressure means the system adjusts to reduce the effect of the change, that is, to reduce the pressure by having fewer gas molecules. During ammonia synthesis, the reactants (i.e., nitrogen and hydrogen) and the product (i.e., ammonia) are in equilibrium, so, to increase the total amount of ammonia formed, the equilibrium should be shifted to product formation by removing ammonia from the reaction mixture as it is produced.

The foregoing expression of reaction equilibrium indicates that one volume of nitrogen and three volumes of hydrogen combine to form two volumes of ammonia. It follows according to Le Chatelier principle that the higher the pressure on the system the larger is the proportion of ammonia at equilibrium, or in other words the equilibrium is shifted towards the smaller volume. On the other hand, the higher the temperature the smaller will be the proportion of ammonia formed. Based on the correlation between the ammonia formed at equilibrium and the effect of temperature and pressure, an ideal ammonia process should be carried out at a temperature as low and at a pressure as high as may be practical and economical.

At 200°C and pressures above 750atm there is an almost 100% conversion of reactants to the ammonia product. Since there are difficulties associated with containing larger amounts of materials at this high pressure, lower pressures of around 200 atm are used industrially.

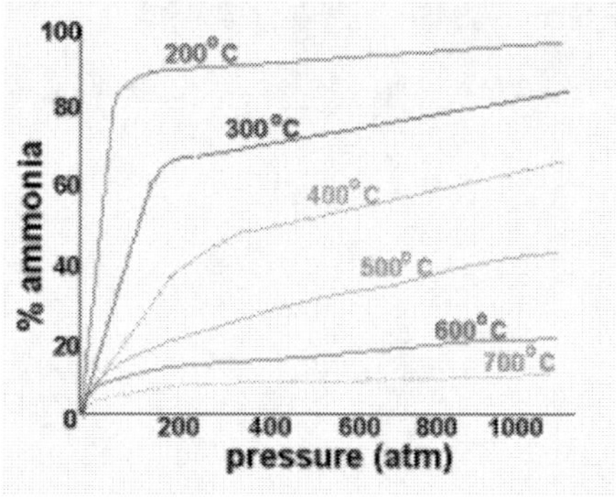

The use of catalyst[5] in ammonia synthesis is extremely important in permitting the reaction to be carried out at a lower temperature than would be practical with no catalyst. A catalyst such as an iron catalyst is used to speed up the reaction by lowering the activation energy so that the N_2 bonds and H_2 bonds can be more readily broken. Increased temperature means more reactant molecules have sufficient energy to overcome the energy barrier to reacting (activation energy) so the reaction is faster at higher temperatures (with lower yield of ammonia).

In commercial plants pressures of around 200atm and temperatures in the range of 400-500°C are use to achieve an acceptable yield of ammonia (10-20%) within an acceptable time period, while costs and safety concerns in the building and during operation of the plant are minimized. The catalysts which are most widely used are based on iron plus small proportions of alumina, calcium, potassium, vanadium, rhutenium etc.

Ammonia Recovery

During ammonia synthesis, the reactants (i.e., nitrogen and hydrogen) and the product (i.e., ammonia) are in equilibrium, so, to increase the total amount of ammonia formed, the equilibrium should be shifted to product formation by removing ammonia from the reaction mixture as it is produced. Traditionally, ammonia NH_3 recovery is carried out by condensation of NH_3 from the reactor off-gases, (cooling the gas mixture to a relatively low temperature e.g., about -5.degC. to about 25.degC.) through a series of heat exchange operations at high pressure usually by refrigeration. This is costly because NH_3 is dilute, and because the stream must be cooled, sometimes under refrigeration. In this temperature range, a two-phase mixture is formed with ammonia being a liquid and nitrogen and hydrogen remaining as gases. The liquefied ammonia is separated from the other components of the mixture.

Subsequently, the remaining un-reacted gas mixture of Nitrogen and hydrogen, inerts and some ammonia (as much as 4-6% un-recovered NH_3), is re-heated to the reaction conditions, recompressed and recycled through the catalyst to the conversion system after the addition of sufficient fresh nitrogen-hydrogen mixture to compensate for the ammonia removed from the system. Even with high-pressure operation, the per pass conversion of synthesis gas to ammonia is 10-20% so that it is necessary to return a large portion of the gas mixture to the converter.

Existing methods recycle at least 4-6% of ammonia back to the reactor which, reducing product formation, (by Le Chetalier's Principle), with a yield of 10-20% of ammonia per pass.

Disadvantages of Current Ammonia Recovery Methods

There are several disadvantages associated with this traditional method of recovery:

- the ammonia separation is often incomplete, (since it would require an infinite amount of energy to lower temperature to completely condense NH3 from the product stream),
- the gas that is recycled to the reactor can include a substantial amount of ammonia. This decreases the efficiency of reactor and ammonia production.
- that a substantial amount of the energy used in this process is related to cooling the gas mixture to separate ammonia by refrigeration
- heavy utilization of mechanical energy for the re-compression of the un-reacted nitrogen and hydrogen gas mixture, some inerts and un-recovered ammonia gas to the pressure conditions of the synthesis reactor.
- energy is needed for the un-reacted gas mixture to be re-heated to the reaction conditions, recompressed and recycled through the catalyst to the conversion system
- the inerts accumulated in the system constitute a burden reducing yield. They are removed mainly through purging, thus wasting valuable reactants (in addition to the fact that they consume appreciable amounts of energy during the compression and refrigeration operations).
- a further complication with the traditional method involves the presence of oxygen-containing species in the synthesis gas mixture during ammonia production. These oxygen-containing species, especially carbon monoxide, can poison the catalyst and reduce the efficiency of ammonia production.

While other methods such as distillation and single-stage flash have been used to separate ammonia from a synthesis gas, however these methods also involve the complexities associated with the separation of certain inert species from the reaction mixture and are energy intensive and expensive

It therefore remains a challenge to provide a method for ammonia production and recovery that provides improved efficiency and reduced cost. In particular, it would be desirable to provide such a method that could cut the cost and avoid the complexity of ammonia condensation. Furthermore, it would be advantageous if the method could be performed without substantial temperature and/or pressure reductions.

Proposed Method and its Advantages

The method we are proposing provides such answers. The proposed recovery concept avoids all of the above limitations. In the proposed method, all of the ammonia produced is recovered and no ammonia is recycled or introduced back into the reactor. The result of this is significantly higher reactor conversion with higher yield of (25-30%) of ammonia.

The new method offers both improved production as well as significant energy savings when applied to U.S. ammonia plants. This saving is from reduced feedstock (natural gas consumption), reduced power requirements for recovery of ammonia through elimination of energy intensive refrigeration system for the recovery of ammonia from the reactor effluent, and reduced in reheating and recompression of recycles streams. The energy from synthesis reactor effluent can be used directly to generate steam for consumption and for export to other parts of the process plant, thus making the system a net energy producer.

The proposed technology can be applied to greenfields. It is also easily applied to retrofit and revamp existing Ammonia plants in the USA. In retrofit, the refrigeration, recompression and reheat units will all be removed and replaced (along with some minor re-piping). The reactor system does not have to be replaced. The new technology achieves 100% recovery of ammonia and no ammonia is recycled back to the reactor. This leads to improved reactor conversion through equilibrium favoring the forward conversion. In a full scale retrofit to take full advantage of the new technology, the absorber, methanation and gas scrubbing units will also be removed. The economic benefits includes: (a) savings in feedstocks (b) significantly reduced power requirements for recovery of ammonia by refrigeration (c) saving in reheating and recompression of recycles. No requirement to reheat or re-compress recycle stream, and (d) reduced capital and operating costs.

Apart from the reduced natural gas feedstock and steam consumption, the proposed method produces pure oxygen as by-product from air-nitrogen purifier. The oxygen can be used directly in furnace to enhance the furnaces' efficiency or oxygen can be exported for other applications.

The scope of the project:
In order to achieve this objective, the project is divided into 4 phases with the following key goals developed:

- Design the ammonia recovery system by pressure swing adsorption
- Investigate to determine and select the demonstration process conditions which support recovery of ammonia (without the need for refrigeration compressor work requirements).
- Design, build, and test integrated pilot scale unit that demonstrate the performance needed for a commercial process.
- Perform economic analysis and compare the expected capital and operating costs of the ammonia process to conventional processes.

The activities in Phase 1 are focused on performing proof of concept to determine viability of the overall system. The key objective for this phase is the successful screening of viable adsorbents, determination of the equilibrium and kinetic data for ammonia recovery section, development of initial PSA Cycle flow-sheet for the ammonia recovery section and initial engineering and economic analysis.

In this presentation, we present initial results including preliminary adsorbent screening and also investigate the potential effects of high temperature and pressure on adsorbent stability in an Ammonia Process.

Acknowledgement:

This project is being sponsored by the US Department of Energy (DOE).

References

[1] Enriching the Earth: Fritz Haber, Carl Bosch, and the Transformation of World Food Production by Vaclav Smil (2001) ISBN 0-262-19449-X
[2] Daniel Charles, Master mind: The Rise and Fall of Fritz Haber, the Nobel Laureate Who Launched the Age of Chemical Warfare (New York: Ecco, 2005), ISBN 0-06-056272-2.
[3] Dietrich Stoltzenberg, Fritz Haber: Chemist, Nobel Laureate, German, Jew: A Biography (Chemical Heritage Foundation, 2005), ISBN 0-941901-24-6.
[4] United States Geological Survey publication
[5] Twygg, Martyn V. (1989). Catalyst Handbook, 2nd Edition, Oxford University Press. ISBN 1-874545-36-7.
[6] Ammonia Catalysis and Manufacture, A. Nielson ed., Springer-Verlag Berlin Heidelberg, NY, N.Y. (1995)

A Novel, Highly Efficient and Economic Purification Process Revolutionizing Purified Terephthalic Acid Production

Randi Wytcherley, GTC Technology, 910 Technology Blvd, Ste F, Bozeman, MT 59718

GTC has developed a revolutionary purification process to improve the purified terephthalic acid (PTA) manufacturing process. GTC is working with the DOE to verify this technology at pilot scale. This development is based on a low-temperature and low-pressure crystallization method, using a highly selective organic solvent. This solvent is non-toxic, environmentally preferred, thermally stable, non-corrosive, non-reactive, and commercially available.

A New Separation Method - Distillative Freezing

Lie-Ding Shiau, Chang Gung University, 259 Wen-Hwa 1st Road, Taoyuan, 333, Taiwan and Chun-Ching Wen, Chemical and Materials Engineering, Chang Gung University, 259 Wen-Hwa 1st Rd, Taoyuan, 333, Taiwan.

The distillative freezing (DF) operation is applied to separate the mixture of the volatile compounds. In principle, the DF process is operated at triple point condition, in which the liquid mixture is simultaneously vaporized and solidified due to the three-phase equilibrium. By lowering temperature and reducing pressure during the operation, the DF process results in the formation of pure solid, and liquid phase and vapor phase of mixtures. The process is continued until the liquid phase is completely eliminated and only the pure solid crystals remain in the feed. Due to the very close boiling points of p-xylene (PX, bp=138.37°C), m-xylene (MX, bp=139.12°C), ethylbenzene (EB, bp=136.19°C), it is very difficult to separate them by conventional distillation. The DF process has been successfully applied to separate PX from a liquid mixture of 10% MX and 90% PX (Shiau et al., 2005). DF is applied in this work to separate and produce the solid PX crystals from the multi-component xylene mixture containing PX, MX, EB and OX. The experimental results indicate that solid PX crystals can be produced and the purity of PX crystals can reach 99.5%~99.9 for the xylene mixture of MX, PX EB and OX with PX ranging from 87 to 90%.

Shiau, L. D.; Wen, C. C.; Lin. B. S., Separation and purification of p-xylene from the mixture of m-xylene and p-xylene by distillative freezing. Ind. Eng. Chem. Res. 2005, 44, P2258-2265.

Mass Transfer Studies on a Novel Rotating Packed Bed (Higee)

Mohammad J. Ansari, Milan Kumar, D.P. Rao, and Nitin Kaistha. Chemical Engineering, Indian Institute of Technology, Kanpur, Room No. H116, Hall 4, IIT Kanpur, Kanpur, India

Keywords: Rotating packed bed; interfacial area; liquid side Mass Transfer Coefficients;

Process Intensification (PI) is a design philosophy that can lead to significant energy, capital, environmental and safety benefits through dramatic reductions in the plant size for a given production objective. Distillation and absorption are two widely used unit operations in industry for separation. The design of these units is governed by gravity force which limits throughput and mass transfer rates resulting in the ubiquitous bulky towers. Ramshaw 1983, first proposed the use of rotating packed beds (RPB) for achieving higher 'g' through centrifugal acceleration with enhancement in mass transfer and throughput rates by 1-2 orders of magnitude and consequent drastic reduction in the column size for the same production objective. The process was aptly termed HIGEE (high gravity).

Research in HIGEE has shown that even as the liquid side volumetric mass transfer coefficient increases by an order-of-magnitude over conventional packed beds, there is little or no enhancement in the gas side volumetric mass transfer coefficient [2]. Rao 2004, proposed a novel RPB design with split-packing to enhance the gas slip velocity for higher gas side mass transfer coefficient. The design of the novel RPB can be seen in Chandra et al, 2005. A photograph of the split packing is shown in Figure 1. Concentric packing rings are attached to the two disks which in turn are attached to two different motor driven shafts. The two disks fit such that there is a small gap between adjacent rings and can thus co-rotate or counter-rotate. The high gas slip velocity in counter-rotation is speculated to result in enhancement in the gas side mass transfer coefficient.

Reddy et al, 2006 showed that the enhancement in volumetric transfer coefficients on the gas side (k_Ga) and liquid side (k_La), respectively, for the novel split packing were in the ranges 35-280 times and 25-250 times compared to packed columns. This work is an extension to separately quantify the interfacial area for mass transfer (a) and the liquid side mass transfer coefficient (k_L) for the novel split-packing RPB design. Danckwert's chemical method is applied for obtaining the interfacial area 'a'. Absorption of CO_2 in aqueous NaOH, a liquid side controlled system, constitutes the experimental system studied. Experiments were done on two types of metal foam packings (specific surface area = 1700 and 2500 m^2/m^3, porosity = 0.9), supplied by RECEMAT International B.V. Holland. Data for CO_2 absorption rates at high NaOH concentration (1 N), where the assumptions of a fast reaction (compared to mass transfer) and pseudo-first order kinetics are valid so that the rate of absorption in independent of k_L, is used for calculating the interfacial area 'a' for mass transfer. Using the values obtained for the interfacial area, the liquid side mass transfer coefficient is calculated from CO_2 absorption rate data at low NaOH concentration (0.05N) where the absorption rate depends on k_L.

The setup (Fig. 2) consists of a rotating packed bed (RPB), a blower, a liquid feed pump, motors and other auxiliary equipments. The blower is used to supply air in the RPB and the centrifugal pump circulates aqueous NaOH through the unit. In the discharge lines of both air and aqueous NaOH, bypass lines are provided to control the flow rate. The flow rates of the liquid are measured by calibrated rotameters. Two types of rotameters are used in the setup. For low liquid flow rates (25.7 and 45.7 ml/s), rotameter having range of 0-50 ml/s was used and for high liquid flow rates (82, 104.57 and 130 ml/s), rotameter having range of 0-300 ml/s is used. Gas flow rate is measured by an orifice meter. The packing RPM is varied from 500 to 2000. CO_2 concentration at the inlet and outlet is measured by "INSPECT AIR CO_2 Meter, Model 8560" by TSI. The inlet and outlet gas CO_2 concentrations are measured to quantify the rate of absorption.

Preliminary experimental results show that the enhancement in liquid side mass transfer coefficient 'k_L' comes out to be 15-35 times that of conventional packed columns. The calculated interfacial area for mass transfer comes out to be about half of the manufacturer specified surface area. Even so, the calculated interfacial area 'a' is 20-40 times that in conventional beds. Detailed results on the variation in the interfacial area and the liquid side mass transfer coefficient with RPM, liquid and gas flow rates and co / counter rotation of the adjacent packing rings will be presented in the conference.

References:

1. Ramshaw, C. (1983), "Higee distillation-an example of process intensification", *The Chemical Engineer,* February, 13-14

2. D.P. Rao, A. Bhowal and P.S. Goswami, "Process Intensification in Rotating Packed Beds (HIGEE): An Appraisal", *Ind.Eng.Chem.Res.*, 43, 1150-1162

3. Gas velocity (m/s)

A.Chandra, P.S. Goswami and D.P. Rao (2005), "Characteristics of flow in a rotating packed bed (HIGEE) with Split Packing", *Ind.Eng.Chem.Res.*, 44, 4051-4060

4. Reddy K.J., A. Gupta, D.P. Rao and O.P. Rama (2006), "Process Intensification in a HIGEE with split packing", *Ind.Eng.Chem.Res.*, 45, 4270-4277

5. M. M. Sharma, P. V. Danckwerts (1970), "Chemical methods of measuring interfacial area and mass transfer coefficients in two fluid systems" *British Chemical Engineering.,* Vol. 15, No. 4, 522-528

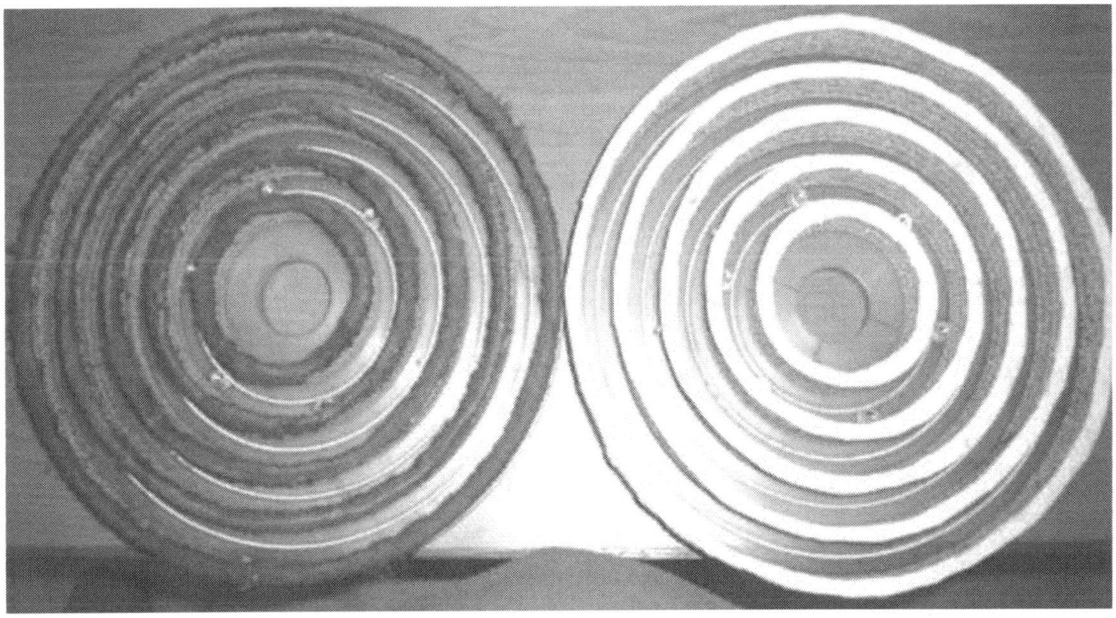

Fig. 1: Photograph of metal foam packing arrangement on the rotor disks

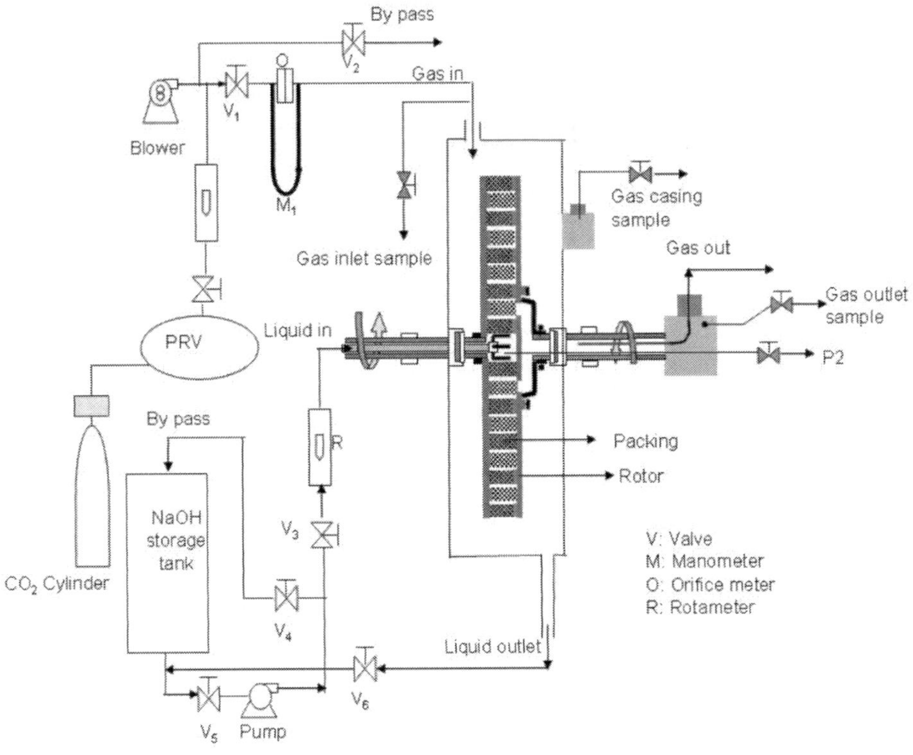

Fig. 2 Schematic diagram of HIGEE unit

A Novel Technology for Virus Vaccine Purification: Modeling and Operation of Continuous Annular Chromatography Unit

Ahmet R. Özdural[1], **Deniz Tanyolaç**[1] and **Nilgün Özdural**[2],

(1)Chemical Engineering Department, Hacettepe University, Beytepe, Ankara, Turkey, (2)Vasat Hazirlama, SAP Institute, Eskisehir Yolu, Ankara, Turkey

Abstract

Recent decades have witnessed a large increase in the technological strategies for vaccine purification. Veterinary vaccines are also experiencing an upsurge in interest, especially for major problem areas such as foot and mouth disease (FMD), which is a devastating disease of live stock. Process economics and process intensification are the key drivers in order to develop a rational purification strategy for FMD vaccines. The continuous annular chromatography (CAC) is a potential and promising downstream process that allows large-scale continuous preparative chromatographic separation and purification of multi-component mixtures. Improvement of the purification of FMD vaccines via continuous annular chromatography (CAC) is the focus of this work. In chromatographic simulation and modeling studies, the complexity of the problem increases if the usually assumed equilibrium-dispersive condition, which neglects all transient resistances, is not invoked. In this work, the non-equilibrium model of Özdural et al [1] is applied to continuous annular chromatography (CAC) modeling studies, so as to understand the contributions of mass transfer resistances on the elution profiles, and thereby providing the efficient and reliable simulation profiles. It was concluded that in order to reach a factual chromatographic column dynamics portrait the inclusion of non-equilibrium constraints in the modeling studies, as separate identities, is essential. In the experimental phase, inactivated and 300K Dalton tangential flow filtered BHK cell culture virus harvest, produced by SAP Institute (FMD vaccine production plant, Ankara, Turkey) is used. It was determined that tangential filtration was not efficient in removing host cell DNA from virus, despite that it is highly competent in regard of other proteins. The virus suspension is further treated with different types of resin filled columns so as to gain an insight of the chromatographic media to be used in CAC which was build upon our design and operation strategies are based on the non-equilibrium model predictions of the model developed in this study. It was demonstrated that careful selection of operation parameters is needed for an efficient separation of virus and host cell DNA, whereas the CAC model predictions might save considerable time for the optimization of chromatographic virus purification strategy. The significance of this work is threefold. First, it delineates the importance of the contributions of mass transfer resistances on the elution profiles of chromatographic separations. Second, use of this new methodology allows us to suggest protocols for system operation parameters and/or scale-up processes of CAC purification systems. Finally, the use of CAC emerged as a novel technology for continuous chromatographic virus purifications.

Keywords: Virus Vaccine Purification, Continuous annular chromatography (CAC); Mass transfer effects; Non-equilibrium adsorption; Convection–dispersion equations.

1. Introduction

Recent decades have witnessed a large increase in the technological strategies for vaccine purification. Veterinary vaccines are also experiencing an upsurge in interest, especially for major problem areas such as foot and mouth disease (FMD), which is a devastating disease of live stock. Process economics and process intensification are the key drivers in order to develop a rational purification strategy for FMD vaccines. Production of FMD virus vaccines via deep suspension cell culture of BHK (baby hamster kidney) cells are first reported by Capstick et al [2], and later this technique is adopted as an industry standard for FMD virus replication. Likewise, SAP Institute (Ankara, Turkey) having a yearly capacity of over 100 million doses of FMD vaccine employs cell culture technique. The virus was purified in a sequence of downstream processing steps comprising of two stage depth filtration, membrane filtration (0.2 micrometer), inactivation, Polyethylene glycol (PEG 6000) treatment, and cross filtration (300K Dalton). In this work, it was determined that tangential filtration was not fully efficient in removing host cell DNA from virus, despite that it is highly competent in regard of other proteins. Surprisingly, despite the importance of cellular DNA clearance for vaccine products, there is a paucity of literature. It is critical to develop scaleable, cost-effective processes which are suitable for commercial manufacturing. In addition, the purification processes must consistently generate virus of high purity to minimize regulatory concern over cellular residuals. It is well known that nucleic acids may adhere to cell derived particles such as viruses. The adhesion may interfere with separation due to agglomeration, change in the size of particle, or change in particle charge, resulting in little if any product being recovered with a given purification scheme. The main advantage of employing chromatography technology for virus purification is that it offers the possibility to selectively and efficiently purify viruses. In this study we have demonstrated that, as a polishing step, chromatographic separation significantly decreases the host cell DNA content in the product fraction.

For large scale separation and purification applications, the major drawback of preparative column chromatography is its discontinuous nature of operation. The continuous annular chromatography (CAC) is a potential and promising downstream process that allows large-scale continuous preparative chromatographic separation and purification. Here, we investigated the viability of the addition of a CAC unit as a final downstream step to the FMD virus purification process.

Fig. 1 illustrates the principle of CAC where the resin phase is packed between the two slowly rotating concentric cylinders building up an annulus. Mobile phase are continuously loaded onto the resin through separate feed and eluent inlets that are located at the top of the resin phase. Keeping the feed and eluent inlet positions fixed, each solute in the feed is forced to flow down through the annulus. The exit angle of individual components are determined by the feed and eluent linear velocities, inlet concentrations, annulus bed height, rotating speed, mobile and stationary phase interaction. Different zones will form helices of characteristic angle which can be collected at various fixed locations around the bottom of the apparatus [3-6]. The concept of CAC was first suggested [7] in 1949 by Dr. A.J.P. Martin. Later several researchers presented the description of the details of construction of an apparatus. The state-of-the-art of the CAC was developed at Oak Ridge National Laboratory, USA [8-12]. In the literature, comprehensive reviews have been published regarding CAC [13, 14]. Several groups of researchers have proposed and solved [15-20] chromatographic columns and packed-bed adsorption models at different levels of complexity and accuracy by making use of equilibrium models, analogy to classical elution chromatography in a conventional fixed-bed. The relationship between time dependent one-dimensional and steady state two-dimensional processes has been discussed in detail by Wankat [21]. Thiele et al. [22] presented a numerical solution methodology for the two dimensional CAC governing equation, based on equilibrium dispersive conditions for multi-component separations. Brozio and Bart [23] concluded to build annular chromatographs with a large radius and a thin annulus in order to minimize the angular dispersion effects.

In one dimensional chromatographic simulation and modeling studies, the complexity of the problem increases if transient resistances are brought into play. Ideal nonlinear chromatography, based on local equilibrium theory, assumes immediate equilibrium between the mobile phase and the stationary phase. Any contributions from hydrodynamic effects or mass transfer phenomena are neglected. However, the assumption of equilibrium conditions prevail throughout a column is not valid in the dynamic state that exists in a chromatographic column [24]. Özdural et al. [1] showed that local equilibrium based frontal chromatographic column simulations become free from errors only at near saturation conditions. In order to account for zone broadening, the well known equilibrium dispersive model of Guiochon [25] is employed by many authors, where instant equilibrium is assumed between the stationary and the mobile phases, and an apparent axial dispersion term is used to describe and understand the band broadening effects of both axial dispersion and the finite rate of the mass transfer kinetics. Increasing influence of mass transfer will result in smoother peaks compared to the ideal shape [15]. However, in chromatographic columns, stationary phase external and internal mass transfer resistances reflect completely different mass transfer mechanisms, as well as axial dispersion. For example stationary phase internal mass transfer resistance is almost immune to that of the changes of mobile phase interstitial velocity, whereas hydrodynamic conditions have a significant effect on both the external mass transfer resistance and axial dispersion terms [26, 27]. It is of interest to note that, during the analysis of adsorption column dynamics, Ruthven [28] indicated that the initial breakthrough time is more sensitive to mass transfer resistance than to axial dispersion whereas the situation is reversed in the saturation region. Hunter and Carta [29] stated that mass transfer within the adsorbent particle is especially important since it is typically the principal cause of band broadening. Lumping all the mass transfer resistances into an apparent dispersion coefficient might lead to erroneous conclusions, due to the fact that the internal and external mass transfer resistances and axial dispersion do not usually respond in a harmony to the changes in column operation parameters. This equilibrium dispersive assumption is even less reliable for the separation of larger, slowly diffusing molecules.

The assessment presented above illustrates that in order to reach a factual chromatographic column dynamics portrait the inclusion of non-equilibrium constraints in the modeling studies, as separate identities, is essential. In this paper the consistency of the application of non-equilibrium model, proposed by Özdural et al. [1] for one dimensional chromatographic simulation, to the two dimensional CAC simulation studies will be explored. The significance of this work is threefold. First, it delineates the importance of the contributions of mass transfer resistances on the elution profiles of chromatographic separations. Second, use of this new methodology allows us to suggest protocols for system operation parameters and/or scale-up processes of CAC purification systems. Finally, the use of CAC emerged as a novel technology for continuous chromatographic virus purifications.

2. Theoretical

The change of interstitial velocity of the liquid stream, v (cm/s), and the liquid concentration gradients in the radial direction of the CAC are considered to be negligible. The mathematical model used in this work considers that single component adsorption takes place. It is assumed that the stationary phase at the CAC annulus is composed of spherical adsorbent particles (radius = r_p cm), it operates under isothermal conditions, the non-linear equilibrium data can be represented by Langmuir equation, and non-equilibrium conditions exist between the adsorbent particle average concentration and the liquid in the void fraction of the annulus. The model is based on a dual resistance model combining external mass transfer and intraparticle transport by homogeneous solid diffusion model (HSDM) and a constant homogeneous solid diffusivity, D_{hs} (cm^2/s) is used throughout the analysis.

In a continuous annular chromatograph (CAC), with referring to Fig. 1 (b), the two spatial directions namely, axial coordinate, x (cm) and angular coordinate, θ (rad) take part in the evaluation of dispersion effects. Eq. (1) gives the governing equation of CAC. This is a well known two dimensional convection-dispersion equation. Independent dispersion coefficients in axial (x) and angular (θ) directions, namely D_x (cm^2/s) and D_θ (rad^2/s) were incorporated into Eq. (1) by many authors [13, 14, 22, 23].

$$\omega\frac{\partial c}{\partial \theta}+v\frac{\partial c}{\partial x}+F\omega\frac{\partial \bar{q}}{\partial \theta}=D_x\frac{\partial^2 c}{\partial x^2}+D_\theta\frac{\partial^2 c}{\partial \theta^2} \tag{1}$$

where

$$F=(1-\varepsilon)/\varepsilon \tag{2}$$
$$\omega=\theta/t \tag{3}$$

c is the angle and bed height dependent bulk liquid concentration (mg/cm^3), $\bar{q}$ is the angle and bed height dependent solid phase particle average concentration (mg/cm^3 solid, including particle pores), ω is the angular velocity (rad/s), F is the phase ratio, ε is the void fraction in CAC bed (-), t is the time (s) and v is the interstitial velocity (cm/s). It is not difficult to derive Eq. (1) by making a mass balance on the control volume shown at Fig. 1 (b) as long as dispersion terms are replaced with axial, D_{xM} and angular, $D_{\theta M}$ molecular diffusivities respectively. However, if it is assumed that dispersion mimics diffusion in the sense that the dispersive fluxes on x and θ dimensions appears to be driven by concentration gradients, and can be expressed using the same mathematical form as Fick's law for diffusive flux, then Eq. (1) is obtained. Wankat [19] provided the following relationship between axial and angular molecular diffusivities in CAC systems, and its derivation is presented elsewhere [30].

$$D_{\theta M}=\frac{D_{xM}}{R_1^2} \tag{4}$$

where the numerical value of R_1 is equal to the annulus outside radius, but having a unit of cm/rad. Eq. (4) reveals that, for large diameter CAC systems, $D_{\theta M}$ value can be neglected in comporison with D_{xM} value, and in return it is likely to come ackross with a similar trend between D_θ and D_x. The conclusions of Byers et al. [3], Brozio and Bart [23] support this hypothesis. Thus, Eq. (1) becomes

$$\omega\frac{\partial c}{\partial \theta}+v\frac{\partial c}{\partial x}+F\omega\frac{\partial \bar{q}}{\partial \theta}=D_x\frac{\partial^2 c}{\partial x^2} \qquad (5)$$

The first and third terms of the L.H.S. of Eq. (5) describe angular coordinate change, and thus the time, of the component in the liquid and solid phases by keeping in the mind that $\omega=\theta/t$. The liquid and solid phase concentrations are related by the linear driving force model [31] which is adapted to CAC systems as following.

$$\omega\frac{\partial \bar{q}}{\partial \theta}=k_f\frac{3}{r_p}(c-c_s^*) \qquad (6)$$

where k_f is the film mass transfer coefficient (cm/s), and c_s^* is the interphase liquid concentration (mg/cm^3). Substitution of Eq. (6) into Eq. (5) gives

$$\omega\frac{\partial c}{\partial \theta}+v\frac{\partial c}{\partial x}+k_f\frac{3}{r_p}F(c-c_s^*)=D_x\frac{\partial^2 c}{\partial x^2} \qquad (7)$$

Regarding the mass transfer resistances, Eq. (7) contains the k_f term only, but as shown in Eq. (8) solid internal resistance is inherently included in the c_s^* term. Eq. (7) contains two dependent variables, c and c_s^*. It can not be solved with its present form. In order to solve Eq. (7), let's express c_s^* in terms of c. The detailed description of the non-equilibrium numerical solution of the chromatographic column version of Eq. (7) is given by Özdural et al. [1]. Yao and Tien [32] presented the batch adsorption version of Eq. (8), by using the LDF approximation.

$$q_s^*=\bar{q}+\frac{Bi}{5}(c-c_s^*) \qquad (8)$$

where q_s^* is the interface solid concentration (mg/cm^3 solid including pore volumes), Bi is the Biot number $(Bi=k_f r_p/D_{hs})$ and D_{hs} is the homogeneous solid diffusivity (cm^2/s) that signify the solid internal resistance. Eq. (9) gives the single component Langmuir adsorption isotherm expression. Common to all chromatographic simulations is the estimation of the equilibrium relationship which describes the concentrations on the stationary phase as a function of the concentrations in the mobile phase. The equilibrium relationship between the stationary and the mobile phase plays a leading role in chromatography; especially for higher concentrations where the equilibrium relationship becomes nonlinear.

For the realistic mass transfer resistances, the widely used instant equilibrium assumption between the solid average concentration ($\bar{q}$) and the bulk liquid concentration (c) is not valid under dynamic conditions [24]. Otherwise, at the classical adsorption isotherm experiments, there would be no need to bring into contact and agitate the solid and liquid phases for prolonged times if the so called equilibrium conditions could instantly be attained between the chromatographic column stationary and mobile phases under dynamic conditions. On the other hand, it is reasonable to assume that the interphase liquid concentration c_s* is in equilibrium with the interphase solid concentration, q_s* since adsorption itself (transfer of solute at the interphase to adsorbed state) is generally very fast. Consequently, the following version of Langmuir adsorption isotherm expression holds for dynamic conditions.

$$q_s* = \frac{q_m K_L c_s*}{1 + K_L c_s*} \tag{9}$$

From Eqs. (8) and (9)

$$c_s* = \frac{-M + \sqrt{M^2 + \frac{4}{K_L}\left[c + \frac{5\bar{q}}{Bi}\right]}}{2} \tag{10}$$

where

$$M = \frac{5q_m}{Bi} - \frac{5\bar{q}}{Bi} + \frac{1}{K_L} - c \tag{11}$$

Once Eq. (10) is substituted into Eq. (7) then c_s* term disappears but a new dependable variable, namely adsorbent particle average concentration, $\bar{q}$ appears. During the numerical solution procedure presented in this work, in order to evaluate $\bar{q}$ value the technique of Özdural et al [1] was employed. Mass balance for the control volume cell, shown at Fig. 1 (b) where there are flux in both the x and θ directions, gives:

$$\bar{q}_{i,j} = \bar{q}_{i,j-1} + \frac{\alpha}{\omega F}(c_{i-1,j} - c_{i+1,j}) + \frac{\gamma}{\omega F}(c_{i-1,j} - 2c_{i,j} + c_{1+1,j}) - \frac{1}{F}(c_{i,j} - c_{i,j-1}) \tag{12}$$

$$\alpha = \frac{v\Delta\theta}{2\Delta x} \tag{13}$$

$$\gamma = \frac{D_x \Delta\theta}{2\Delta x^2} \tag{14}$$

Fig. 2 illustrates i and j indexes that are used as x and θ dimensions panel indexes respectively. Since Eq. (12) allows the numerical calculation of $\bar{q}_{i,j}$ value from one step former θ dimension panel value ie., $q_{i,j-1}$. Therefore in Eq. (15) only one independent variable ie. bulk liquid concentration, c is left to be calculated.

$$\omega\frac{\partial c}{\partial\theta}+v\frac{\partial c}{\partial x}+k_f\frac{3}{r_p}F\left(c-\frac{-M+\sqrt{M^2+\frac{4}{K_L}\left[c+\frac{5\bar{q}}{Bi}\right]}}{2}\right)=D_x\frac{\partial^2 c}{\partial x^2} \tag{15}$$

Under non-linear conditions such as the present case, there are no analytical solutions for Eq. (15). Solutions must be calculated numerically with the following periodic boundary conditions, where θ_f is the feed angle. Number of j panels corresponding to the feed angle $=\theta_f/\Delta\theta$.

Boundary conditions:
$\theta=0$ *and* 2π then $c(x,0)=c(x,2\pi)$
At the top of the annulus (x = 0)
for $0\le\theta\le\theta_f$ then $c(0,\theta)=c_0$; otherwise $c(0,\theta)=0$
At the bottom of the annulus (x = L)
$\partial[c(L,\theta)]/\partial[x]=0$

Since explicit schemes may suffer from stability limits finite differences with implicit schemes method is used. $\partial c/\partial x$ and $\partial^2 c/\partial x^2$ terms are evaluated by central difference approximation, and $\partial c/\partial\theta$ term is approximated by forward difference with reference to the grid network shown in Fig. 2. Instead of applying above given B.C. at x=L, we approximated the $c(x,\theta)|_{x=L}$ values at i= nx+1 by linear extrapolation of $c(x,\theta)$ values at i = nx-1 and i = nx. As shown in Fig. 2, if we have a grid of n+1 spatial points for x direction, then at angle j+1 there are n+1 unknown nodal values. We can assemble the set of n+1 equations. By solving this equation system, we determine c_1, c_2, ... c_{n+1} at angle step j+1 from c_1, c_2, ... c_{n+1} at angle step j. Özdural et al. employed a similar solution methodology to frontal chromatography and packed bed catalytic reactors [1, 33]. After discretizing the transport equation the set of n+1 algebraic equations to be solved simultaneously are as follows. Eq. (17) shows that during the solution procedure $c_{i,j}$* values are required. These values are calculated through Eq. (10) by employing $\bar{q}_{i,j}$ values given by Eq. (12)

i = 1 *(top of annulus)*

$$c_{1,j+1}=c_{1,j}=c_0 \text{ or } 0 \text{ (boundary condition dictates)} \tag{16}$$

1< i < nx+1

$$-(\alpha+\gamma)c_{i-1,j+1}+(\omega+2\gamma)c_{i,j+1}+(\alpha-\gamma)c_{i+1,j+1}=(\omega-\beta k)c_{i,j}+\beta k c_{i,j}{}^* \tag{17}$$

where $\beta=(3/r_p)k_f F$ and $k=\Delta\theta$

i = nx+1 (bottom of annulus)

$$c_{nx-1,j+1}-2c_{nx,j+1}+c_{nx+1,j+1}=0 \tag{18}$$

It might be interesting to note that the present approach of the solution of convection-diffusion adsorption equation obviates the solely trailed complicated procedure described in the literature [33-35], where solution of "coupled" partial differential equation (PDE) systems is required. The present methodology employs a single PDE by crediting the particle phase concentration at a specified bed and angular location through the use of mass balance expressions which concurrently takes into account the accumulation and axial dispersion effects.

3. Simulation results

The influence of the parameters on the CAC model output was studied for single component adsorption and elution. The algorithm presented is found to be sufficiently fast and easily converges for the non-linear equilibrium isotherm studied under non-equilibrium conditions.

The configuration and model operating parameters for the CAC system are summarized in Table 1. In Figs. 3 and 4 CAC concentrations outlet profiles are presented for both the solid and eluent phases. In order to observe the influence of the feed concentration model simulation is repeated using parameter values of Table 1 but the value of c_0 is doubled. Fig 3 (a) indicates that position of solid concentration peak maxima shifts and it appears earlier for high feed concentration. This can be attributed to the mass transfer resistances encountered during desorption stage. Since the present model is based on non-equilibrium conditions the high solute concentration retained in the solid can not be easily flushed out as assumed by equilibrium models. Thus for high feed concentrations the solid concentration peak maxima appears earlier. On the other hand, Fig. 3 (b) shows that the concentration of eluent at the outlet is higher for high feed concentrations but peak maxima position shift is much less numerable. This phenomenon is again the direct consequence of mass transfer limitations, where for low feed concentrations there is a second trend of a smooth eluent concentration rise. It might be interesting to indicate that such a behavior might occur for certain experimental conditions, but can not be simulated in equilibrium model solutions since it incorrectly predicts a narrower mass transfer zone than actually is present. Fig 4 indicates the effect of stationary phase internal mass transfer resistance on solid and eluent outlet concentrations. Contrary to the generally used equilibrium dispersive model where instant equilibrium is assumed between the stationary and the mobile phases, and an apparent axial dispersion term is used to describe and understand the band broadening effects, these figures clearly illustrate the importance of the inclusion mass transfer resistances on model studies.

4. Experimental

Baby hamster kidney $BHK_{21}An_{30}M_{17}S_{15}$ cell growth and virus propagation were conducted in suspension culture in stirred tank bioreactors. FMD seed virus, SOV-36 was used for these experiments. Prior to any of the downstream processes, unclarified virus culture broth were inactivated by adding BEI (binaryetyleneimine). The virus was purified in a sequence of downstream processing steps comprising of two stage depth filtration (Sartorius, Germany). In the first stage Sartorius sartoclear CP3 (11 μm) filter media was used, whereas in the second stage Sartorius sartoclear F7MP (1 μm) was employed as the depth filter media. Then the virus culture is filtered through Sartorius sartobran P-IV (0.2 μm) cellulose acetate membranes filter and subsequently concentrated with 300K Dalton cross filter system (Sartorius). In order to precipitate dsDNA, and other contaminating proteins the concentrated virus culture is treated with PEG 6000 solution. The final vaccine is prepared in oil adjuvant according to the propriety of SAP Institute.

Starting from the above described SAP vaccine, adjuvant free virus extract was prepared for the estimation of DNA content. This allows us to make a comparison between the various FMD vaccines available on the market, regarding their DNA contents. DNA quantification was performed in a 10x10 mm cuvette using the Quant-iT PicoGreen dsDNA reagent (Molecular Probes, Inc., USA). For sample dilution Tris–EDTA (TE) buffer according to the manufacturer's protocol was used. The assay was calibrated with lambda DNA standard, provided in the Quant-iT PicoGreen dsDNA kit in the linear range between 1 and 1000 ng/ml using the 200-fold TE-buffer diluted reagent (PicoGreen working solution) and TE buffer for dilution purposes. For the determination of dsDNA in adjuvant free FMD vaccine, to 100 μl sample solution, 900 μl TE buffer and 1000 μl of the 200-fold TE-buffer diluted PicoGreen working solution were added respectively and incubated at room temperature for 5 min. Afterwards the fluorescence (excitation: 475 nm, emission: 515 nm) intensity was measured using a Fluorometer (Turner Designs Model 13095-02 USA). Despite the fact that it changes from batch to batch, the host cell DNA concentration in the adjuvant free FMD vaccine extract were quantitated within the range of 0.5 - 1 μg/ml.

In order to gain an insight of the number of main protein structures, sample adjuvant free vaccine extract is injected to NUCLEOSIL 100-5 C_{18} column of the HPLC apparatus (Shimadzu, Japan). Fig. 5 shows the chromatogram obtained with isocratic elution where UV detector (280 nm) is used. The experimental conditions were as follows. Eluent: 50 % TFA,+ 50 % acetonitrile. Flow rate: 0.6 ml/min. Injected sample volume: 5μl. Fig 5 shows two distinct but close peaks that might cause a difficult setting during continuous chromatographic separations with CAC.

A second column with a matrix of 6% highly cross-linked spherical agarose, mean particle size 90 μm (HiPrep 16/10 Q XL, GE Healthcare Bio-Sciences, Sweden) is used as an ion exchange chromatography column so as to increase the time lag between the appearances of the dsDNA peak and the virus peak. Fig. 6 shows the chromatogram obtained with HiPrep16/10 Q XL column. The experimental conditions were as follows: eluent B: 20 mM Tris-HCL pH 7.0, eluent A: 20 mM Tris-HCL -0.5 M NaCl pH 7.0, gradient from 0 100 % A in 25 min, then 100 % A, flow rate: 0.4 ml/min, injected sample volume: 50μl, detection: UV, 280 nm. Fig. 6 illustrates the viability of preparative chromatographic separation of virus from the host cell DNA. For the continuous chromatographic purification, experimental studies with a locally manufactured CAC system are started, where CAC annulus thickness is 6 mm, outside diameter is12 cm, and net resin packing height 24 is cm. As a sample and eluent pumps, FMI model QSY-1CSC (Fluid Metering Inc., USA) and Reglo-Z (Ismatec, Switzerland) pumps are used respectively. The CAC apparatus is filled with Q Sepharose XL virus licensed (GE Healthcare Bio-Sciences, Sweden) ion exchange adsorbent.

5. Conclusions

The present work proposes a non-equilibrium algorithm for the numerical solution of convection–dispersion adsorption equations. A theoretical study has been presented for non-linear non-equilibrium modeling of continuous annular chromatography. In the literature, the neglect of mass transfer resistances by equilibrium dispersive approach considered the theoretical optimum which could be achieved for the chromatographic simulation studies. In reality, however, mass transfer effects could not be eliminated from the separation process.

It was concluded that tangential filtration and precipitation with PEG was not efficient in removing host cell DNA from virus, despite that it is highly competent in regard of other proteins. On the other hand it was demonstrated that with the addition of ion exchange chromatography unit (adsorbent: highly cross-linked, bead formed 6% agarose matrix) efficient separation of host cell DNA is possible. Another significance of this work is that, the process can be operated under continuous conditions with careful selection of the operation parameters of Continuous Annular Chromatograph (CAC).

Furthermore human virus vaccines, such as human influenza vaccines are conventionally produced in the chicken eggs. From there they are purified, e.g. by a combination of centrifugation, ultrafiltration and subsequently continuous sucrose gradient ultracentrifugation [37]. This labor-intensive production method requires a large quantity of eggs, approximately one egg per trivalent dose, which can lead to a supply shortage especially in case of a pandemic outbreak [38]. In addition to the scalability problem egg produced vaccines can cause allergic reactions induced by egg proteins. Such concerns have led to the development of cell culture based production processes [39-41]. The Continuous Annular Chromatograph (CAC) might also find applications in cell culture based human virus vaccines for purification purposes.

6. Nomenclature

Bi	Biot number ($k_f.r_p/Ds$)
c	bulk liquid concentration (mg/cm^3)
c_0	feed liquid concentration (mg/cm^3)
c_s*	interphase liquid concentration (mg/cm^3)
D_x	axial dispersion coefficient (cm^2/s)
D_{xM}	axial molecular diffusivity (cm^2/s)
D_θ	angular dispersion coefficient (rad^2/s)
$D_{\theta M}$	angular molecular diffusivity (rad^2/s)
D_{hs}	homogeneous solid diffusivity (cm^2/s)
i	x panel index used in numerical solution
j	θ panel index used in numerical solution
k_f	film mass transfer coefficient (cm/s)
F	phase ratio equal to $(1 - \varepsilon)/\varepsilon$
K_L	constant in Langmuir isotherm (cm^3/mg)
L	annulus height (cm)
m	annulus thickness (cm)

$\bar{q}$ adsorbent particle average concentration (mg/cm^3 solid including particle pore volume)
q_s * interface solid concentration (mg/cm^3 solid including particle pore volume)
q_m Langmuir isotherm maximum adsorption capacity (mg/cm^3 solid including particle pore volume)
Q_e Eluent flow rate (cm^3/s)
Q_f Feed flow rate, (cm^3/s)
r CAC radial coordinate (cm)
r_p adsorbent particle average radius (cm)
R annulus outside radius (cm)
R_1 parameter in Eq. (4). Its numerical value is the same with R but having a unit of cm/rad.
t time (s)
u superficial velocity in axial direction (cm/s)
v interstitial velocity in axial direction (cm/s)
x CAC axial distance (cm)

Greek letters

α definiton is given by Eq. (13)
β definiton is given by Eq. (18)
ε void fraction in CAC bed
γ definiton is given by Eq. (14)
η . length of arc (cm) that corresponds to θ angle for a specified radius.
θ CAC angular position (rad)
ρ adsorbent particle density (g/cm^3)
ω angular velocity (rad/s)

References

[1] A.R. Özdural, A. Alkan, P.J.A.M. Kerkhof, J. Chromatogr. A. 1041 (2004) 77.
[2] P.B. Capstick, A.J. Garland, W.G. Chapman, R.C. Masters, Nature, 205 (1965)1135.
[3] C.H.Byers, W.G. Sisson, J.P. DeCarli II, G. Carta, Biotechnol Prog. 6 (1990) 13.
[4] H.J. Bart, R.C. Messenböcl, C.H. Byers, A. Prior, J. Wolfgang, Chem. Eng. Process. 35 (1996) 459.
[5] K. Reissner, A. Prior, J. Wolfgang, H.J. Bart, C.H. Byers, J. Chromatogr. A. 763 (1997) 49.
[6] A. Uretschlager, A. Jungbauer, J. Chromatogr. A. 890 (2000) 53.
[7] A.J.P. Martin, Discussions of the Faraday Society. 7 (1949) 332.
[8] R.M. Canon, W.G. Sisson, J Liq. Chromatogr. 1 (1978) 427.
[9] R.J. Torres, C.S. Chang, H.A.Epstein, Fundamental Studies in Continuous Annular Chromatography, (Oak Ridge National Lab., TN), Report: ORNL/MIT-329, Sep 1981.
[10] J.M. Begovich, W.G. Sisson, AIChE J. 30 (1984) 705.
[11] W.G. Sisson, J.M. Begovich, C.H. Byers, C.D. Scott, Application of Continuous Annular Chromatography to Size-Exclusion Separations, (Oak Ridge National Lab., TN), Report: CONF-870802-10, 1987.

[12] J.P. DeCarli, G. Carta, C.H. Byers, Advanced Techniques for Energy-efficient Industrial-scale Continuous Chromatography, (Oak Ridge National Lab., TN), Report: ORNL/TM-11282, 1989.
[13] A. Uretschlager, A. Jungbauer, Bioproc. Biosyst. Eng. 25 (2002) 129.
[14] F. Hilbrig, R. Freitag, J. Chromatogr. B. 790 (2003) 1.
[15] R. Zenhausern, D.W.T. Rippin, Comput. Chem. Eng. 22 (1998) 259.
[16] G. Dünnebier, K. U. Klatt, Chem. Eng. Sci. 55 (2000) 373.
[17] K. Kaczmarski, D. Antos, H. Sajonz, P. Sajonz, G. Guiochon, J. Chromatogr. A. 925 (2001) 1.
[18] A. Rajendran, O. Kräuchi, M. Mazzotti, M. Morbidelli, J. Chromatogr. A. 1092 (2005) 149.
[19] M.A. Hashim, K.H. Chu, Sep. Purif. Technol., 53 (2007) 189.
[20] T.M. Pabst, G. Carta, J. Chromatogr. A., 1142 (2007) 19.
[21] P.C. Wankat, AIChE J. 23 (1977) 859.
[22] A. Thiele, T. Falk, L. Tobiska, A. Seidel-Morgenstern, Comput. Chem. Eng. 25 (2001) 1089.
[23] J. Brozio H.J. Bart, Chem. Eng. Technol. 27 (2004) 962.
[24] D.A. Skoog, D.M. West, F. J. Holler, S.R. Crouch, Fundamentals of Analytical Chemistry (8th Ed). Brooks Cole, Belmont, 2003.
[25] G. Guiochon, S. Golshan Shirazi, A.M. Kati. Fundamentals of Preparative and Nonlinear Chromatography. Academic Press, Boston, 1994.
[26] C.J. Geankoplis. Transport Processes and Separation Process Principles (4th Ed). Prentice Hall, New Jersey, 2003.
[27] S.C. Foo, R.G. Rice, AIChE J. 21 (1975) 1149.
[28] D.M. Ruthven. Principles of Adsorption and Adsorption Processes. John Wiley & Sons, New York, 1984.
[29] A.K. Hunter, G. Carta, J. Chromatogr. A. 971 (2002) 105.
[30] T. Koyuncu, Non-Equilibrium Modeling of Continuous Annular Chromatography. M.Sc. Thesis (in Turkish). Hacettepe University, Ankara, 2006.
[31] E. Glueckauf, J.I. Coates, J. Chem. Soc. (1947) 1315.
[32] C. Yao, C. Tien, Chem. Eng. Sci. 47 (1992) 457.
[33] A.R. Özdural, A. Alkan, C. Webb, Chem. Eng. J. 118 (2006) 17.
[34] A.I. Liapis, D.W.T. Rippin, Chem. Eng. Sci. 33 (1978) 593.
[35] K. Kaczmarski, M. Mazzotti, G. Storti, M. Morbidelli, Comput. Chem. Eng. 21 (1997) 641.
[36] G. Xiu, P. Li, A.E. Rodrigues, Chem. Eng. J., 95 (2003) 83.
[37] M.R. Ikizler, P.F. Wright, Vaccine, 20 (2002)1393.
[38] L. Opitz, J. Salaklang, H. Büttner, U. Reichl, M.W. Wolff, Vaccine, 25 (2007) 939.
[39] http://www.usatoday.com/money/industries/health/drugs/2004-10-29-flu-uastx.htm
[40] http://198.73.159.217/sanofipasteur/ImageServlet?imageCode=17707&siteCode=AVPI_US
[41] http://www.medicalnewstoday.com/medicalnews.php?newsid=46600

Figure 1. Schematic view of CAC. (a) Operating principles, (b) Lay-out for mathematical analysis.

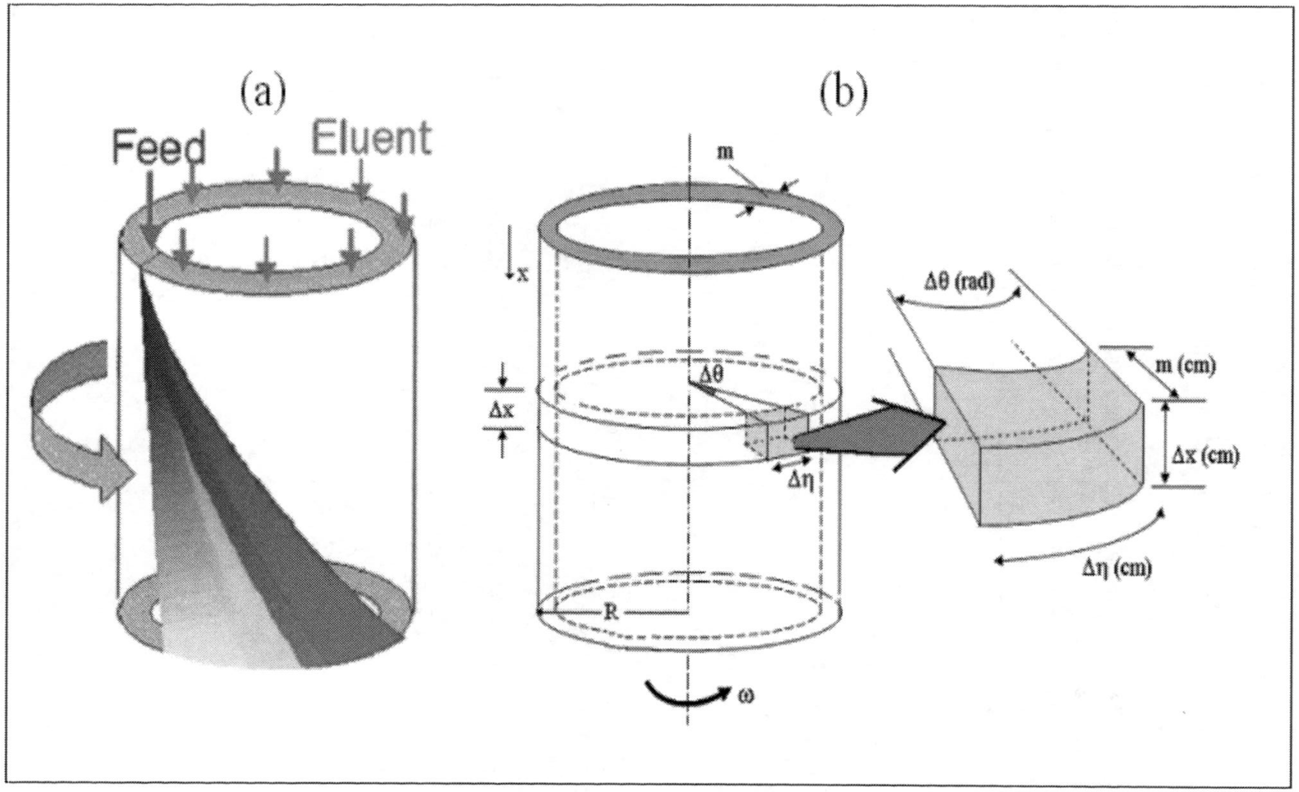

Figure 2. CAC grid display for x and θ dimension panel indexes.

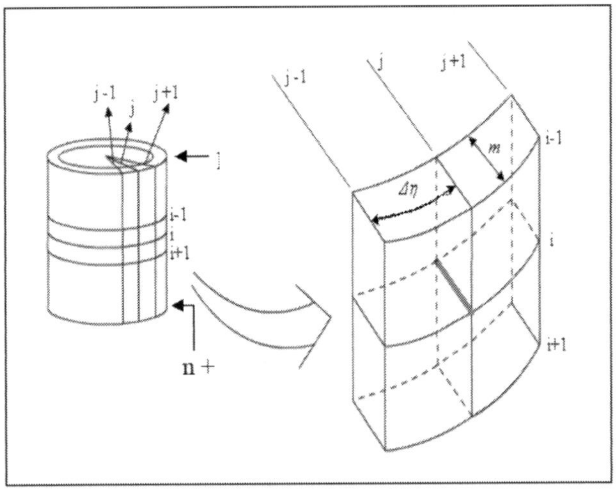

Table 1. The configuration and pivot operating parameters of CAC for simulation studies

Properties	
Average particle diameter, r_p (μm)	100
Bed voidage, ε (-)	0.40
Annulus height, L (cm)	28
Annulus outside radius, R (cm)	10
Annulus thickness, m (cm)	1
Feed angle (annulus opening for feed), θ_f (rad)	π/100
Langmuir isotherm q_m value (mg/cm^3 media)	10
Langmuir isotherm K_L value (cm^3/mg)	0.25
Homogeneous solid diffusivity, D_{hs} (cm^2/s)	2.5×10^{-8}
External film mass transfer coefficient, k_f (cm/s)	1×10^{-3}
Axial dispersion coefficient, D_x (cm^2/s)	1×10^{-3}
Angular velocity, ω (rad/h)	4π
Feed inlet concentration (mg/cm^3)	50
Feed flow rate, Q_f (cm^3/min)	0.58
Eluent flow rate, Q_e (cm^3/min)	118.98

Figure 3. **Prediction of the effect of feed concentration on the stationary and mobile phase concentration profiles of CAC.**

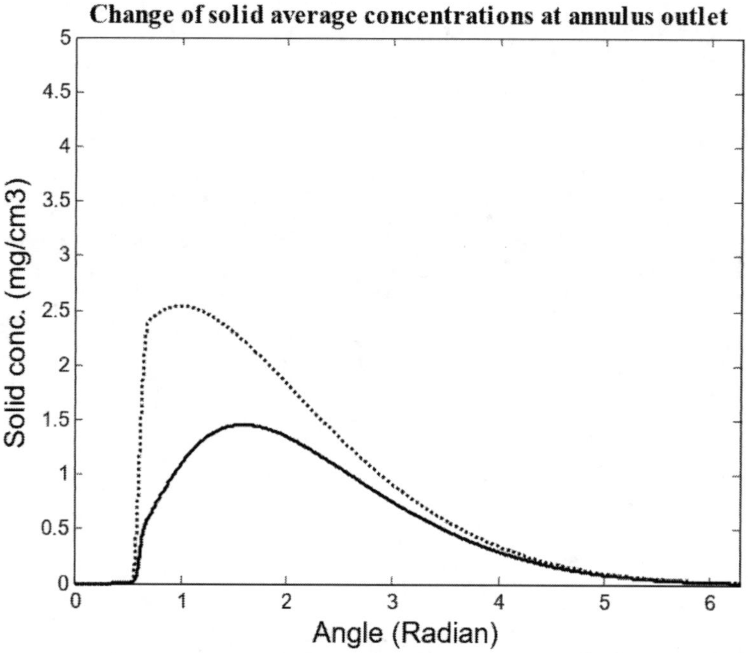

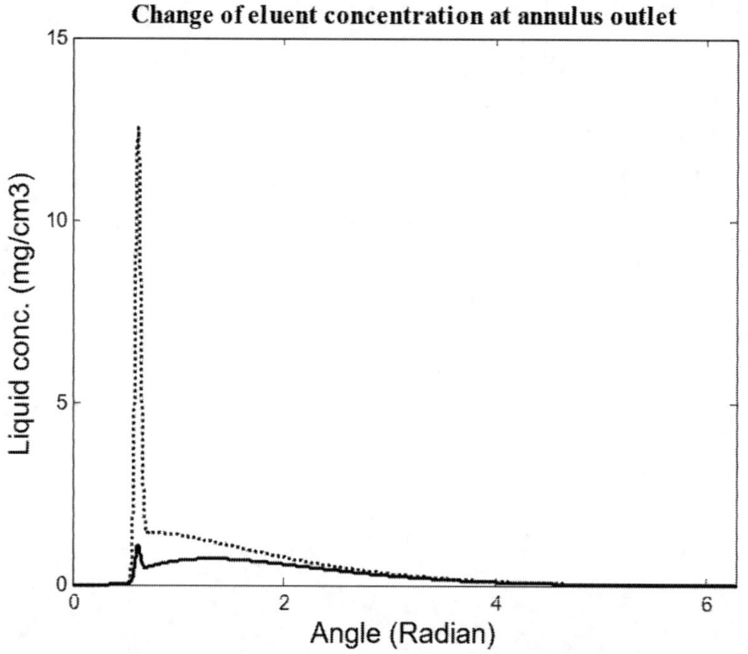

Figure 4. **Prediction of the effect stationary phase internal mass transfer resistances on the stationary and mobile phase concentration profiles of CAC.**

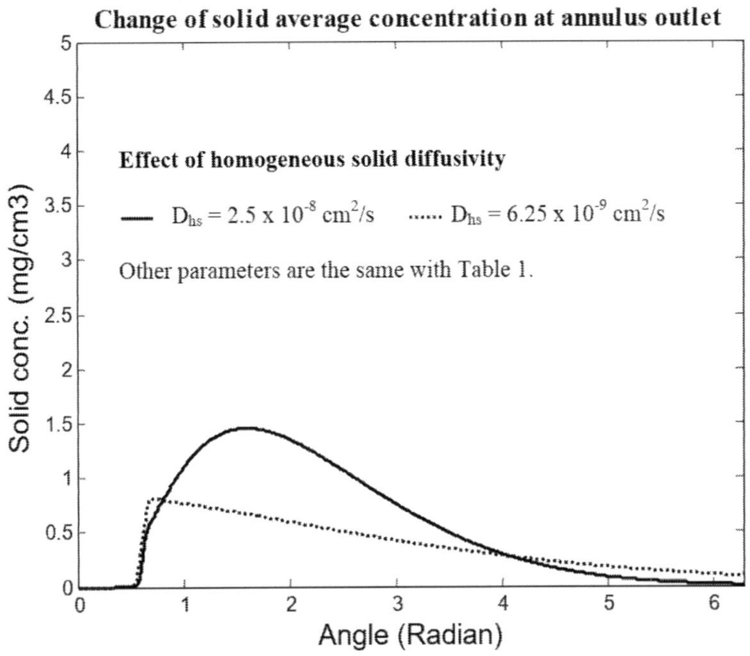

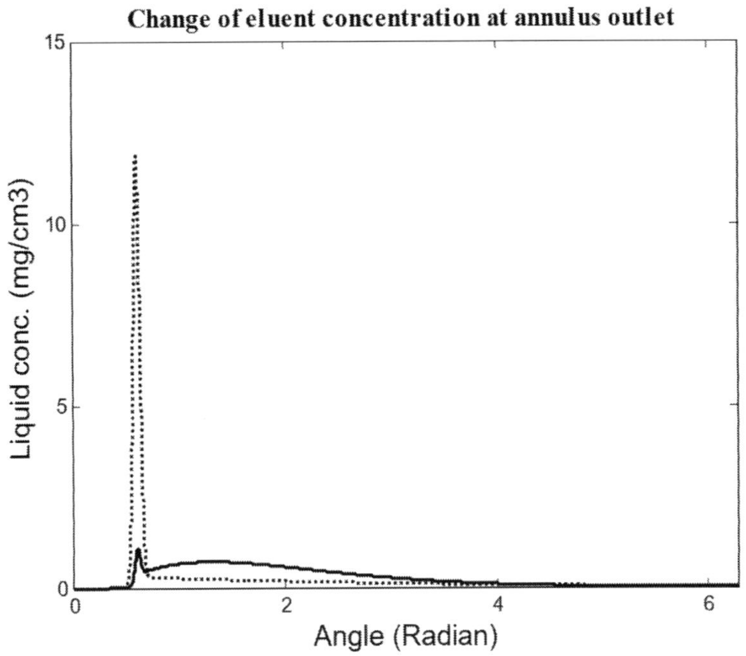

Figure 5. **Chromatogram obtained from D vaccine extract. Column:NUCLEOSIL 100-5 C_{18}, 250 x 4.6 mm ID, isocratic elution: 50 % TFA +50% acetonitrile, flow rate: 0.6 ml/min, injected sample volume:5µl, detection: UV, 280 nm**

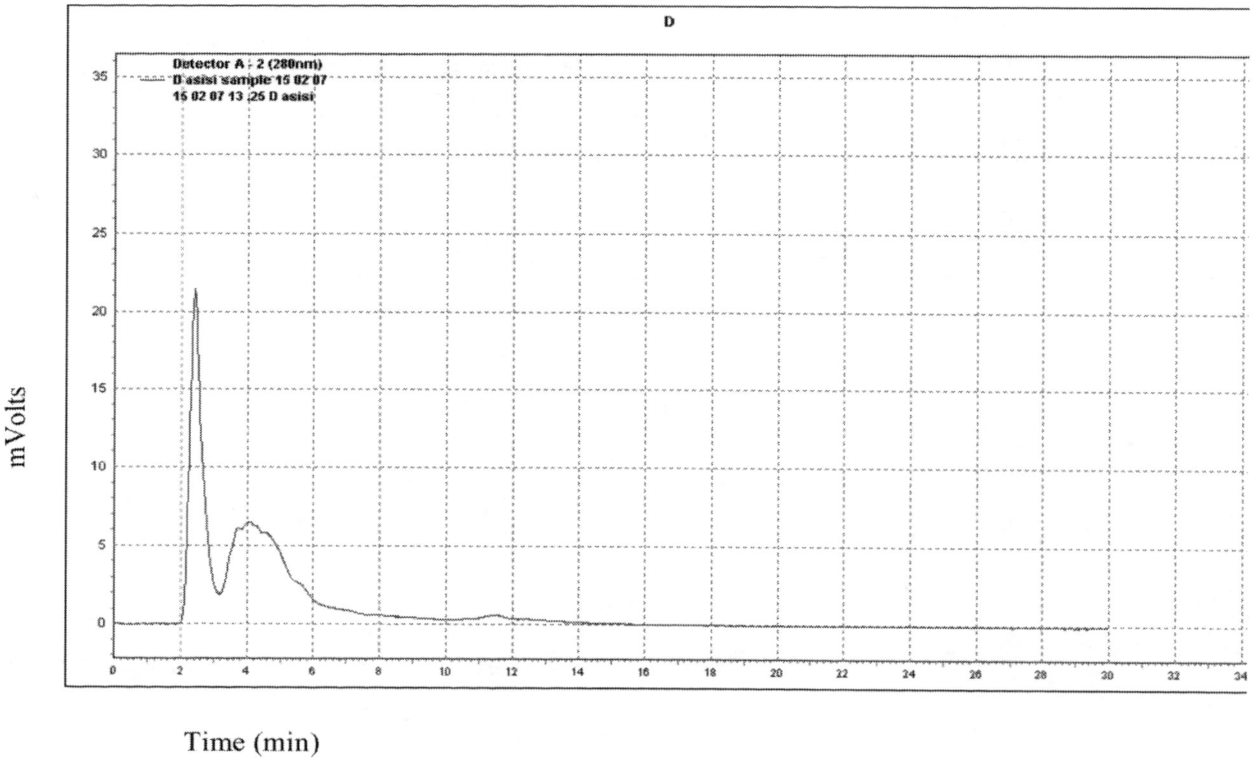

Figure 6. **Chromatogram obtained from D vaccine extract. Column: HiPrep 16/10 Q XL ™, 100 x 16 mm ID, eluent B: 20 mM Tris-HCL pH 7.0, eluent A: 20 mM Tris-HCL -0.5 M NaCl pH 7.0, gradient from 0 100 % A in 25 min, then 100 % A, flow rate: 0.4 ml/min, injected sample volume: 50µl, detection: UV, 280 nm**

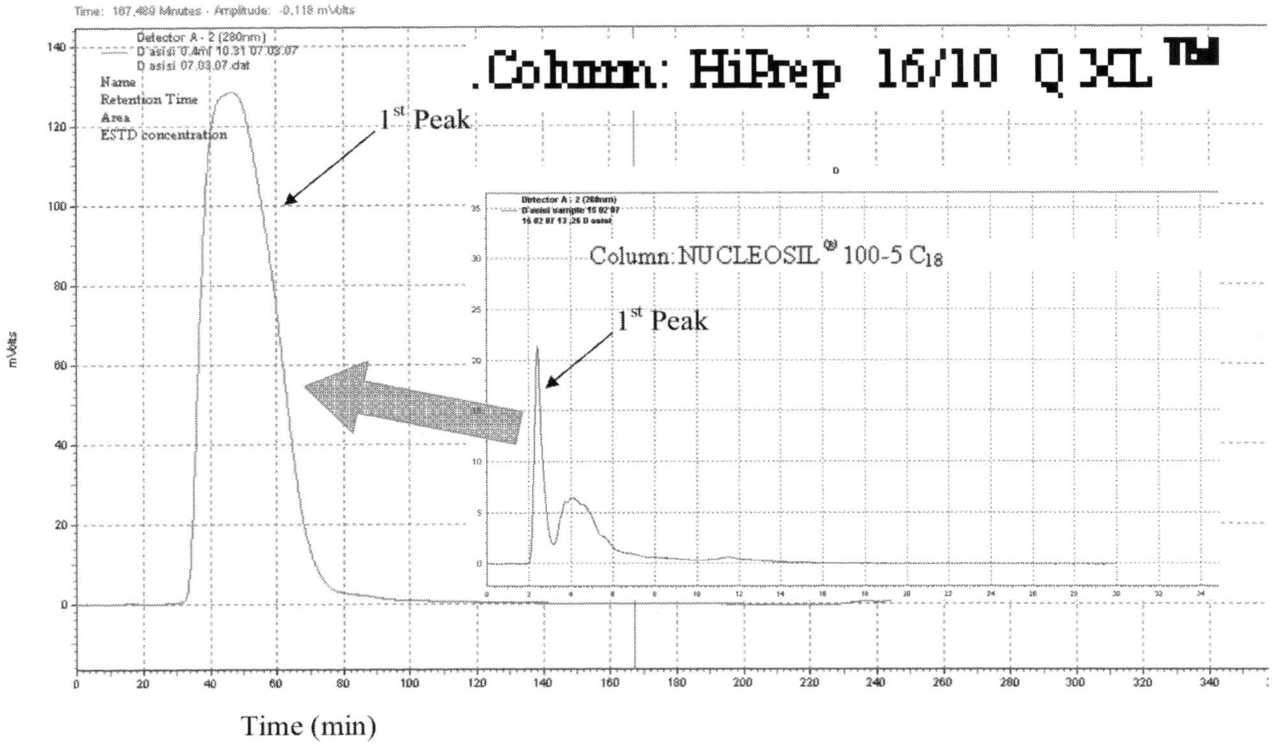

Estimation of Deuterium Recovery from the Separation of Water-Isotope Mixture by Thermal Diffusion

Ho-Ming Yeh, Tamkang University, Department of Chemical and Materials Engineering, Tamsui, 251, Taiwan

ABSTRACT Nuclear energy perhaps will be supplied by thermonuclear fusion with deuterium as the nuclear fuel, instead of fission reactors, in the distant future. Thermal diffusion is a well-established method for separating isotopes, especially for hydrogen isotopes because of the large ratio in molecular weights. It was great achievement of Clusius and Dickel (Naturwiss. 1938;26:546-52) to point out that a horizontal temperature gradient produces not only thermal diffusion in the direction of the temperature gradient, but also natural convection of the fluid upward near the hot surface and downward near the cold surface. These convection currents produce a cascading effect analogous to the multistage effect of a countercurrent extraction or distillation, and as a result a considerably greater separation may be obtained. The first complete presentation of the separation theory of thermal diffusion was that of Furry and co-workers (Phys. Rev 1939;55:1083-95) The separation of water-isotope mixture (H2O-HDO-D2O) in a batch-type thermal diffusion column was done in the previous work (Chem. Eng. Sci 1984:39:1277-82). The separation theory, as well as the time histrories of the concentrions of HDO (C2) and D2O (C3), were derived with the assumption of local equilibrium (D2O+H2O°ê2HDO) within the column. In present study, the pseudo-product form of concentration, , was simplified, and the recovery of deuterium (D) is thus estimated by calculating the concentration difference of D at the bottom (CD,B) and top (CD,T) of the column as (CD,B-CD,T)=[(4/20)C3,B+(2/19)C2,B]-[(4/20)C3,T+(2/19)C2,T]. It is found that the experimental results are in fair agreement with prediction values.

Development of Highly Selective Oxidation Catalysts by Atomic Layer Deposition

Jeffrey W. Elam[1], Michael J. Pellin[2], Larry A. Curtiss[2], Hau H. Wang[2], Gregory K. Krumdick[1], Joseph A. Libera[1], Gerry W. Zajac[3], Steven A. Cohen[3], and Peter C. Stair[4]. (1) Energy Systems Division, Argonne National Laboratory, 9700 S Cass Ave, Argonne, IL 60439, (2) Materials Science Division, Argonne National Laboratory, 9700 S Cass Ave, Argonne, IL 60439, (3) Ineos Technologies, 150 W Warrenville Rd, Naperville, IL 60563, (4) Department of Chemistry, Northwestern University and Chemistry Division, Argonne National Laboratory, 9700 S Cass Ave, Argonne, IL 60439

The goal of the project is to develop new techniques for the manufacture of catalysts by atomic layer deposition (ALD) for the selective oxidative dehydrogenation of alkanes. The specific project objective is to demonstrate the catalyst manufacturing techniques by developing, characterizing, testing, and evaluating new catalysts for the direct ammoxidation of propane to acrylonitrile on existing catalytic substrates in conventional reactors and in single-pass catalytic reactors utilizing catalytic membranes. Our work has focused so far on the preparation of the ALD catalysts comprised of the oxides of vanadium, niobium and other transition metals. In our initial efforts, we have developed ALD chemistries for depositing these oxide materials onto flat substrates to facilitate analysis of the thin film layers. Next, we applied the ALD oxide materials onto high surface area silica gel powders as well as anodic aluminum oxide (AAO) membranes. We have characterized these catalysts using a variety of techniques including scanning electron microscopy, energy dispersive analysis of X-rays, X-ray photoelectron spectroscopy, X-ray diffraction, and X-ray fluorescence. Catalytic testing of these ALD materials is beginning.

Web Page: www.es.anl.gov/Energy_systems/Atomic_Layer_Deposition/Index.html

Effect on Fluidization of Cold and Hot Gas Injection into Fluidized Beds

Sibashis S. Banerjee, Millennium Inorganic Chemicals, Glen Burnie, MD, USA
Timothy O'Hern, Sandia National Laboratories, Albuquerque, NM, USA
Dale Snider, Arena-flow, LLC, Albuquerque, NM, USA
Ken Williams, Arena-flow, LLC, Albuquerque, NM, USA
Rutton D. Patel, ExxonMobil Research & Engineering VA,USA
Alvin U. Chen, ExxonMobil Research & Engineering, VA, USA

Introduction

Industrial fluidized beds often have jets at different temperatures injected into them. These jets may carry liquids, solid feed material, or gaseous reactants. The effects of these jets have been studied experimentally under laboratory conditions. There have been very few computational studies involving injection of jets into fluidized beds and their comparison with experimental data. The purpose of this paper is to compare the predictions of the MP-PIC method involving jets injected into fluidized beds.

Experimental Apparatus

The experimental apparatus is outlined by Chen and Weinstein[1,2]. The bed is a rectangular bed 0.38 m wide and 0.15 m wide. The height is 1.22 m. The bed material is FCC catalyst with a density of 1450 kg/m^3. The mean particle size is 59 μm. The other properties are outlined in Table 1. The bed was fluidized with air room temperature at velocities ranging from 0.03 m/s to 0.21 m/s. Air at different temperatures was injected using either a 0.64 cm diameter nozzle or a 1.27 cm diameter nozzle located at 0.31 m above the base of the bed. For jet air at room temperature the nozzle velocities ranged from 23 m/s to 69 m/s. For the heated case, the velocities ranged from 15.24 m/s to 39.63 m/s. The solid volume fractions at various sections of the bed were measured using X-rays. In addition, for the heated jet case, the temperature distribution was also measured using a rack of thermocouples.

Table 1 Experimental Conditions[1]

Mean Particle Size	59 μm
Sphericity	1
Particle Density	1450 kg/m^3
Minimum Fluidization Velocity	0.9 cm/s
Void Fraction at Minimum Fluidization	0.48

Numerical Method

The commercial code Barracuda™ was used for the calculations. The numerical method has been outlined previously (Snider, 2001). This software was developed specifically for fast and accurate computation of gas-solid flows and is used for diverse industrial applications. The numerical method treats the particle phase as both a continuum and as discrete particles. Particle properties are mapped to and from and Eulerian grid. On this grid, continuum derivatives that treat the particle phase as a fluid are evaluated and mapped back to the individual particles. This allows for accurate handling of particle loadings ranging from dilute to dense with a distribution of densities and sizes. Since the numerical method is detailed elsewhere it is not described here.

Results

A total of four cases were calculated for this paper. The first two were isothermal cases with a jet diameter of 1.27cm with a jet velocity of 23 m/s and 46 m/s. For the non-isothermal case the jet velocities were kept the same but temperatures of 500K and 1000K were used. The time-averaged volume fractions after the introduction of the jets were compared to the base case without a jet. This gives an extent of penetration of the jet into the bed. The time-averaged thermal profiles give the extent of thermal penetration and temperature diffusion in the bed. Fig. 1 shows the particles colored by instantaneous solids volume fraction before the introduction of the jet. Fig. 2 shows the particles colored by instantaneous solids volume fraction after the introduction of the jet at 23 m/s. The jet penetrates less than 5 cm in the horizontal direction consistent with the experimental observations. The bubbles formed by the jet climb up near the wall adjacent to the jet. The rest of the bed is relatively unaffected.

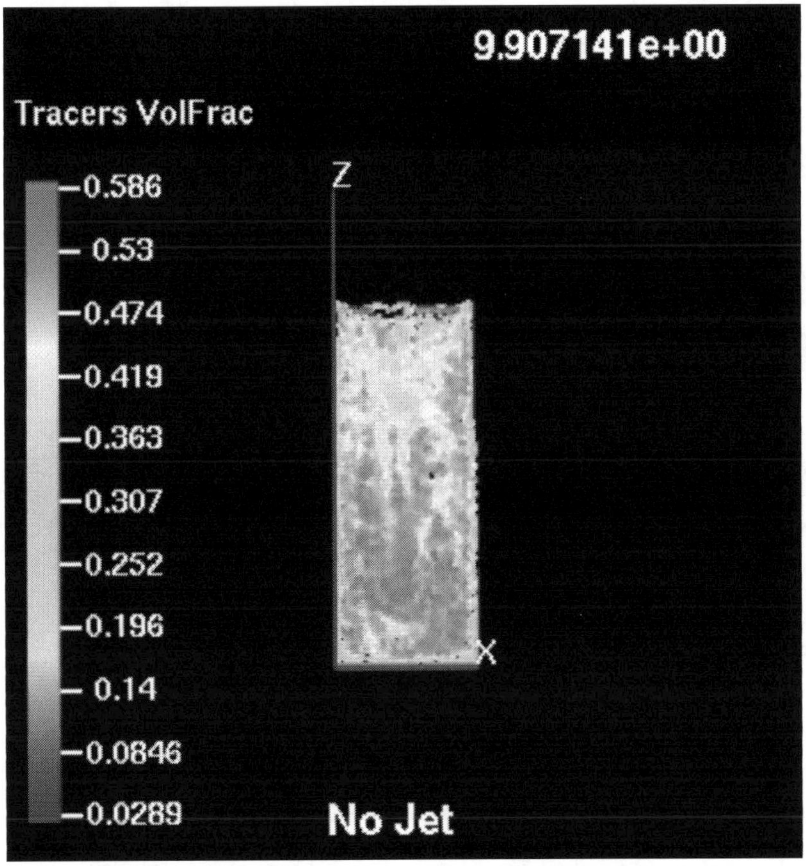

Fig. 1 Instantaneous Particle Volume Fraction (before jet)

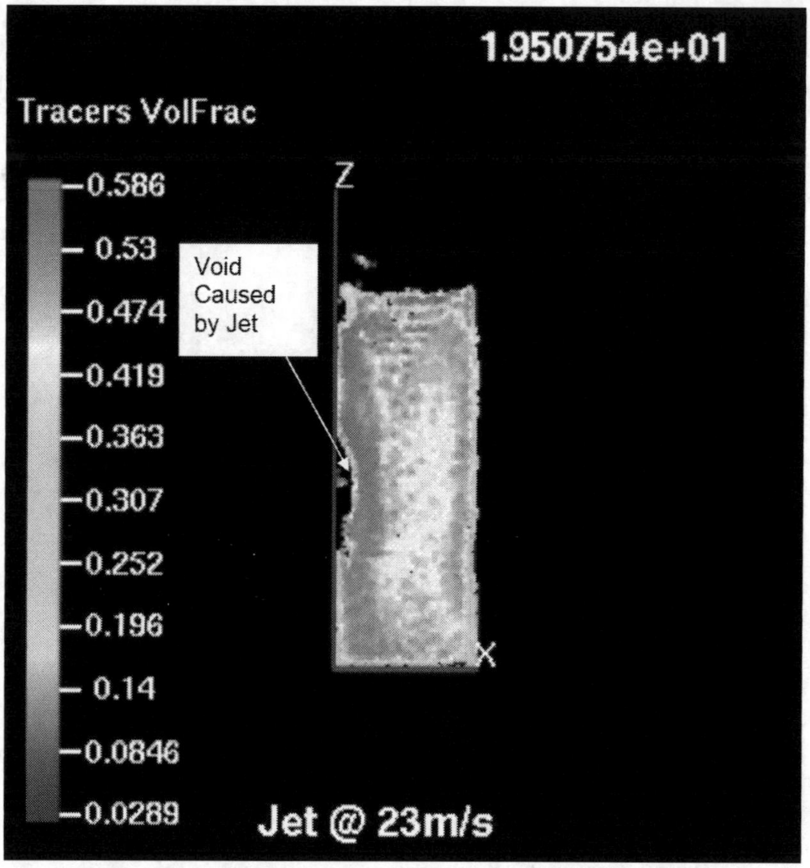

Fig. 2 Instantaneous Particle Volume Fraction (after jet)

Acknowledgements

The authors gratefully acknowledge the support of the Department of Energy Industrial Technologies Program. This program is funded under award number DE-FC36-04GO14153.

References

1. Chen, Libin and Weinstein, Herbert (1993), "Shape and Extent of the Void Formed by a Horizontal Jet in a Fluidized Bed," AIChE Journal, Vol. 39, No. 12, pp. 1901-1909.
2. Chen, Libin and Weinstein, Herbert (1997), "Temperature Distribution Around Heated Horizontal Jet in Fluidized Bed," AIChE Journal, Vol. 43, No. 9, pp. 2373-2375.
3. Snider, D. M. (2001), "An Incompressible Three-Dimensional Multiphase Particle-in-Cell Model for Dense Particle Flows," Journal of Computational Physics, 170, pp.523-549.

Direct Ethoxylation of Butyl Acrylate with Narrow Adduct Distribution

Yun Fang* Yong-an Shen Shi-run Ji Yong-mei Xia

(School of Chemical and Material Engineering, Southern Yangtze University,1800 Lihu Avenue, Wuxi, Jiangsu 214122, China)

Abstract One-pot production of ethylene glycol monobutyl ether acrylate was achieved by direct ethoxylation of butyl acrylate, using a Al-Mg composite oxide catalyst. Mass percentage of ethylene glycol monobutyl ether acrylate in the crude product was 40.3% after 30min reaction at 150-155 °C when a molar ratio of 2:1(butyl acrylate to ethelene oxide) was employed. The catalyst activity was 0.35gEO/(g cat • min). 4-oxo-2,2,6,6-tetramethyl piperdine-1-oxyl was found the most effective inhibitor to prevent self-polymerization of butyl acrylate during the ethoxylation.

Keywords: Ethoxylation; butyl acrylate; ring opening;composite oxide; polymerization; ethylene glycol monobutyl ether acrylate

Introduction

Ethylene glycol monoalkyl ether acrylate is a useful monomer providing a better and adjustable hydrophility than methyl acrylate in the synthesis of functional polymers[1-4]. So far, the production of (poly)ethylene glycol monoalkyl ether acrylate has been conducted in a two-step procedure, including etherealization of the alcohol, and the subsequent esterification between ethylene glycol monoalkyl ether and acrylic acid or the transesterification between ethylene glycol monoalkyl ether and methyl acrylate. The synthesis of ethoxylated saturated aliphatic esters has been studied well[5-9], including the one-step synthesis of ethoxylated oil and fats[8]; We also reported the synthesis of oligo-ethylene glycol ethyl ether acetate[9], in which an Al-Mg composite oxide modified with slight amount of Fe, Ni or some other metal (all were group VIII metals) was applied as the catalyst. However, the ethoxylation of acrylate has not been well documented. Since the reaction temperature of ethoxylation of acrylate would be higher than that of ethoxylation of saturated aliphatic esters due to the conjugative effect of the double bonds; therefore the most difficulty of the synthesis is to avoid self-polymerization of acrylate and get a narrow distribution of the addition number if polyethylene glycol monoalkyl ether acrylate is the target product. Yurug[10] used an Mg-Al-based mixed oxide catalyst to synthesize oligo-ethylene glycol monoalkyl ether acrylate in the presence of phenothiazine as a polymerization inhibitor, in which the amount of the catalyst was about 2.5%-5% of the weight of starting material acrylate, and the reaction time was 4-5hr; however, they did not mentioned about the synthesis of polyethylene glycol monoalkyl ether acrylate. We reported herein an one-pot production of ethylene glycol monobutyl ether acrylate by using 2,2,6,6-tetramethyl-1-piperidinyloxy (TMPO) as a high temperature inhibitor, and an Al-Mg composite oxide modified with slight amount of Fe as the catalyst.

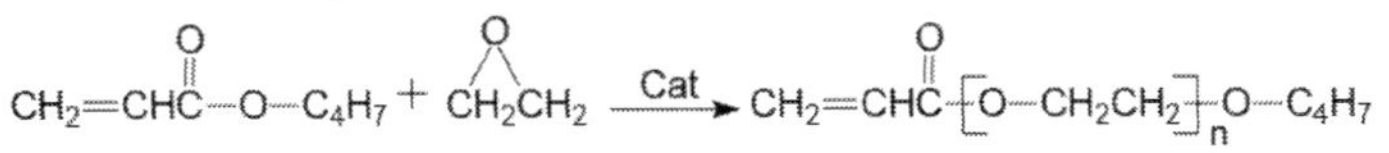

where n＝1 （EGBEA), 2 (DEGBEA), 3 (TEGBEA) .

Scheme 1. One-step production of ethylene glycol monobutyl ether acrylate

* Correspondence to: Yun Fang, School of Chemical and Material Engineering, Southern Yangtze University,1800 Lihu Avenue, Wuxi, Jiangsu 214122, China. E-mail: yunfang@126.com

Experimental

Chemicals

Ethylene oxide was purchased from Sinopec Yangzi Petrochemical Company Ltd. All other chemicals were of reagent grade and used as received.

Analysis

Extra EA was removed by distillation from the reaction mixture before GC or GC-MS analysis. FT-IR, GC-MS and ESI-MS (ES^+) were used to identify the compositions of crude product. GC-MS was equipped with a PEG-30 column, GC temperature program was initially at 50 °C for 2 min, and increased up to 200 °C at 10 °C/min, and then held at 200 °C for 15 min. Mass fraction (w_i) of each component in the reaction mixture was used to indicate the product selectivity. Ethyl heptanoate was applied as an internal standard.

Preparation of the catalyst

Al-Mg composite oxide catalyst with a molar ratio of Al/Mg of 0.6 was prepared as following: 40 g NaOH and 24 g Na_2CO_3 were dissolved in 400 mL deionised water to form solution A. Solution B was prepared by dissolving 67.53g $Al(NO_3)_3 \cdot 9H_2O$, 76.2g $Mg(NO_3)_2 \cdot 6H_2O$, and 2g $Fe(NO_3)_3 \cdot 6H_2O$ in 600mL deionised water. Then solution B was added into solution A in 30 min under vigorous stirring. The resulting solution was of pH 9-10; it was then kept at 70°C for18h, and filtered thereafter. The precipitate was dried and calcined at 500°C for 5.5h.

Preparation of oligo-ethylene glycol monobutyl ether acrylate

100 g butyl acrylate, 1 g inhibitor and 1 g catalyst were added into the 500 mL autoclave under nitrogen atmosphere. Desired amount of EO was introduced into the reactor with a constant feeding rate at the set temperature. The reaction was hold for 15 min upon finishing the feeding. A transparent liquid product was obtained thereafter.

Preparation of polyethylene glycol (10) monobutyl ether acrylate

30g Butyl acrylate,1.1g Al-Mg catalyst,80g petroleum ether (as a dilutor for Butyl acrylate, bp: 90-120℃) and TMPO (1g) were added to a 500mL autoclave equipped with a thermometer, a stirrer, and gas inlet. The reactor was then closed and deaired with nitrogen. The reaction mixture was heated to 150-155 ℃. Ethylene oxide was then added drop wisely to the mixture in 1hr, where n(BA) : n(EO) = 1 : 10 (mol/mol). The reaction was allowed to for another 15min upon finish the feeding of ethylene oxide.

Results and Discussion

Screen of the inhibitors

We have reported the one-step synthesis of oligo-ethylene glycol ethyl ether acetate using same catalyst in detail[9]. However, as indicated in forehead, the conjugative effect of the double bonds in acrylate requires a higher ethoxylation temperature than acetate does, and this enhances the self-polymerization of acrylate further. To prevent the self-polymerization of acrylate, some commercial available inhibitors used in high temperature such as Cupric N,N-dibutyl dithiocarbamate(AI-6lR), 2,2,6,6-tetramethyl-1-piperidinyloxy (TMPO), phenothiazine (PZ) were compared for ethoxylation of butyl acrylate in different temperature. As shown in Table 1, AI-6lR and TMPO presented potential ability to inhibit the self-polymerization. Among which, TMPO was found the most effective.

Table 1. Evaluation of the inhibitors for the synthesis [a]

Reaction temperature /℃	Inhibitors used in the ethoxylation of butyl acrylate		
	PZ	AI-6IR	TMPO
120-125	no self-polymerization was found after cooling 2h	no self-polymerization was found after cooling 6h	no self-polymerization was found after cooling 6h
140-145	self-polymerization was found after cooling 1h	Partially self-polymerization was found after cooling 4h	no self-polymerization was found after cooling 4h
150-155	self-polymerization was found after cooling 1h	Partially self-polymerization was found after cooling 2h	no self-polymerization was found after cooling 3h

a: Reaction conditions: 0.35gEO/(gCAT • min), 1.5hr, n(BA) ∶ n(EO) = 2 ∶ 1(mol/mol)

Influence of temperature on the ethoxylation of butyl acrylate

The influence of temperature on the reaction is showed in Table 2. Reactions conducted above 160 ℃ were accompanied with self-polymerization of butyl acrylate more or less depending on the reaction temperature after 1.5hr reaction.

Table 2. Influence of temperature on the reaction[a]

Run	Reaction temeprature /℃	Inhibitor	Result
1	140-145	AI-6IR	Partially self-polymerized, acrylate was ethoxylated
2	140-145	TMPO	No self-polymerization, no ethoxylated acrylate
3[b]	150-155	TMPO	No self-polymerization, acrylate was ethoxylated

a. Reaction conditions: 0.35gEO/(gCAT • min), 1.5hr, n (BA) ∶ n(EO) = 2 ∶ 1(mol/mol)

b. The product contains (w/w): w(EGBEA)=40.3 %、w(DEGBEA)=24.7 %、w(TEGBEA)=19.7 %、w(TEEGBEA)=5.1%

Preparation of polyethylene glycol (10) monobutyl ether acrylate

Polyethylene glycol (10) monobutyl ether acrylate is a special surface-active monomer providing the built polymer better hydrophility. The product was analysized with ESI-MS (ES^+) as shown in figure1. The mass-to-charge ratio $(M+Na)^+$of 415.8、459.9、503.9、548.0、592.1、636.2、680.2、724.3、768.4、812.5、856.6、900.6、944.8、988.8 represented the produced polyethylene glycol monobutyl ether acrylate with addition number of 6-19, respectively; of which, the content of products with addition number from 8 to 12 is 62.4% (mol/mol) or 58.3 %(w/w) in the reaction mixture.

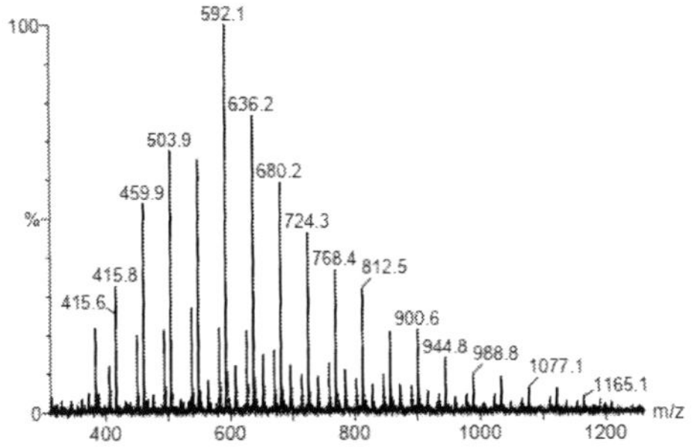

30g Butyl acrylate, n(BA) ∶ n(EO) = 1 ∶ 10 (mol/mol),1.1g Al-Mg catalyst,80g petroleum ether (bp: 90-120 ℃), 1g TMPO;150-155 ℃.

Figure 1 ESI-MS spectra of polyethylene glycol (10) monobutyl ether acrylate

Conclusions

One-pot production of ethylene glycol monobutyl ether acrylate can be achieved by direct ethoxylation of butyl acrylate, using an Al-Mg composite oxide catalyst. 4-oxo-2,2,6,6-tetramethyl piperdine-1-oxyl was found the most effective inhibitor to prevent self-polymerization of butyl acrylate during the ethoxylation. The ethoxylation of butyl acrylate does not occur below 150 ^{0}C, but came up with the self-polymerization of butyl acrylate above 160^{0}C. Mass percentage of ethylene glycol monobutyl ether acrylate in the crude product was 40.3 % after 30min reaction at 150-155 ^{0}C, while polyethylene glycol (10) monobutyl acrylate with the addition numbers from 8 to 12 was obtained of 62.4% (mol/mol) in the reaction mixture.

Acknowledgment

Financial assistances from Ministry of Education of China for Key Research Projects of Science and Technology (2002-78) and Program for New Century Excellent Talents in University by Ministry of Education of China are gratefully acknowledged.

References

1. Brian C. Anderson, Surya K. Mallapragada. Synthesis and characterization of injectable, water-soluble copolymers of teritiary amine methacrylates and poly(ethylene glycol) containing methacrylates. *Biomaterials*, 2002, 23:4345-4352.
2. Stefan Dreher, Neustadt, Joachim Pakusch Speyer, Mario Sandor Worms *et al*. Cement compositions comprising redispersible polymer powders. US.6489381(2002).
3. Fu Zhang, E.T. Kang, K.G. Neoh et al. Surface modification of stainless steel by grafting of poly(ethylene glycol) for reduction in protein adsorption. *Biomaterials*, 2001, 22:1541-1548.
4. Hsin-Cheng Chiu, Chorng-Shyan Chern, Cheng-Kang Lee *et al*. Synthesis and characterization of amphiphilic poly(ethylene glycol) graft coplymers and their potential application as drug carriers. *Polymer*, 1998, 39:1609-1616.
5. Scholtz HJ, Suehler H, Quack JM, Schuler W, Trautmann M. Process for the Preparation of Carboxylic Acid Esters of Alkylene Glycol Ethers and Their Utilization. EP 89105357.1(1989).
6. Alejski K, Bialowas E, Hreczuch W, Trathnigg B, Szymanowski J. Oxyethylation of Fatty acid methyl esters Molar ratio and temperature effects Pressure drop modeling. *Ind. Eng. Chem. Res.* 2003, 42: 2924-2933.
7. Hama I, Okamoto T, Hidai E, and Yamada K, Direct ethoxylation of fatty methyl ester over Al-Mg composite oxide catalyst. *Journal of the American Oil Chemists' Society* 1997,74: 19-24.
8. Yang CY, Fang Y and Chen J, Study of direct alkoxylation of natural fats & oils and emulsifying properties of the products. *Fine Chemicals* 2004, 21(4): 262-265.
9. Yong-mei Xia, Yong-an Shen, Shi-run Ji, Yun Fang, Yi-you Wang, Bing Tan. One-step production of alkoxylated ethyl acetate with narrow adduct distribution. *Journal of Chemical Technology and Biotechnology*, 2007, 82: 47-50.
10. Keiji Yurug, Takafumi Kubo. Process for producing alpha, beta-unsaturated carboxylic acid esters and catalyst for use in such process. EP 1123915A1(2001).

On the Use of Co-Solvent in Counter Current Multistage Continuous Liquid Extraction Operations

Dr. Kal Renganathan Sharma PE

Adjunct Professor, Department of Chemical Engineering (MC 110)
University of Illinois at Chicago
202 Chemical Engineering Building
810 South Clinton Avenue, Chicago, IL 60607
Tel: (312) 996 3424

Extended Abstract

The common denominator between the liquid extraction and solid leaching process is the solvent used. In *liquid extraction* bulk separations are effected from liquid mixtures and in *solid leaching* the bulk separations are carried out from solid mixtures. Liquid extraction is used in areas where it is difficult to separate the components otherwise. Some *solvent loss* during a continuous process with recycle will add to the *cost of the separation*. The original liquid mixture is called the *feed* and upon extraction becomes the *raffinate*. The *extract* refers to the portion of the product that contains the desired component or solute in cases where the feed mixture was a solution. If the desired component to be separated is a gas then the operation of separation would be *stripping*.

Liquid extraction is selected after consideration has been given to distillation and absorption. Liquid liquid extraction is central in some industries. It is used in petrochemical, pharmaceutical, metallurgical and energy industries. The general principles shown for distillation, humidification is applicable here also. Extraction is used to separate metal ions that are non-volatile. It is used in flavor and fragrance industry where the compounds that need to be handled can decompose at conditions of distillation. Solvent extraction can be used to separate azeotropes as discusses in section 4.3. Typical solvents used and the application used for is shown in Table 1.0. Lubricants are made from specific fractions of the crude oil after refining operations.

The process of liquid extraction is used in the preparation of most drugs. Synthetic drugs produced by fermentation are separated by liquid extraction. The production of bacitracin involves partitioning from the bacterial growth medium into butanol. Gelatin preparation involves the use of warm water as solvent after the skin and bone collagen is treated with lime or dilute acid. Extraction of the other proteins and polypeptides involves use of solvent on the minced animal tissue. Penicilin is produced using fermentation. The suspended microbes in the beer are removed by filtration. The clarified beer is then extracted with an alkyl acetate such as amyl acetate. Pencilin undergoes four transfers before being crystallized into the final product. Supercritcal CO_2

is used for extraction. One example is in coal cleaning. Liquefied coal can be further extracted into useful fractions. The different industries liquid extraction is used are as follows;

1. Chemical

o Washing of acids/bases, polar compounds from organics

2. Pharmaceuticals

- Recovery of active materials from fermentation broths
- Purification of vitamin products

3. Effluent Treatment
   o Recovery of phenol, DMF, DMAC
   o Recovery of acetic acid

4. Polymer Processing
   o Recovery of caprolactam

   o Separation of catalyst

5. Petroleum
   o Lube oil quality improvement
   o Separation of aromatics/aliphatics (BTX)

6. Petrochemicals
   o Separation of olefins/paraffin
   o Separation of structural isomers

7. Food Industry
   o Decaffeination of coffee and tea
   o Separation of essential oils (flavors and fragrances)

8. Metals Industry
   o Copper production
   o Recovery of rare earth elements

9. Inorganic Chemicals
   o Purification of phosphoric acid

10.0 Nuclear Industry
   o Purification of uranium

A typical extraction process is shown in Figure 8.1. The solute is given by B instead of C and solvent denoted by C instead of B. Since most investment and operating costs are associated with the solvent recovery steps, it is very important to consider and study this aspect when designing the entire process. In a typical extraction process about 3% of the operating cost is in the extractor, with the remaining 97% in solvent recovery. Therefore, it is extremely important to consider the solvent recovery aspects early in the project since they play such an important role in overall process economics. The feed, raffinate, extract and solvent can be recognized in the process flow diagram.

Table 1.0 Applications of Liquid Extraction in Industry

No.	**Solvent used**	**Application**	**Feed**	**Industry**
1.	Amyl acetate, Methylene Chloride	Penicilin Extraction	Filtered fermentation beer	Pharmaceuticals and food
2.	Warm water	Gelatin manufacture	Skin and bone collagen	biotechnology
3.	Propane, Hexane	Refining fats and oils	Soyabeans	Pharmaceuticals and food
4.	Hydroxylamine in kerosene	Concentrating copper for electrowining	Acidic leach liquors	Mining and Metallurgy
5.	Tertiaryamines In kerosene	Uranium and rare earth separations	Acidic leach liquors	Nuclear Power Industry
6.	Glycols, furfural,cresol, liquid SO_2	Dewaxing lubricating oils	Crube lube oil	Petroleum
7.	Glycols, sulfolane	Higher octane aromatic fuels	Aliphatic-aromatic mixtures	Petroleum
8.	Aqueous Base	Desulfurization	Sour distillates	Petroleum
9.	Aqueous Copper Complex	Butadiene/Butene separation	Incompletely Dehydrogenated Feed	Plastic
10	Toluene	Monomer purification	Impure caprolactum	Plastic
11	Ethylbenzene	Continuous Polymerization of Polystyrene - Recycle	Styrene, initiator	Plastic
12	Methylethyl ketone	Continuous Polymerization of ABS - Recycle	Styrene, Acrylonitrile, Polybutadiene	Plastic
13	Phenol	Recovery and	Polycarbonate	Plastic

		Recycle in Chemical Process	manufacture	
14	Ethyl Acetate/Butyl Acetate	Separation of Hydrogen Bonded Organics from Water	Waste water	Water Treatment
15	Hydrocarbon/ aqueous alcohol	Orange Oil Cinnamon Oil Peppermint Oil Lemon Oil	Essential Oil	Food, Fragrance Industry
16	Water	Organics	Organic/base	Organic Chemicals

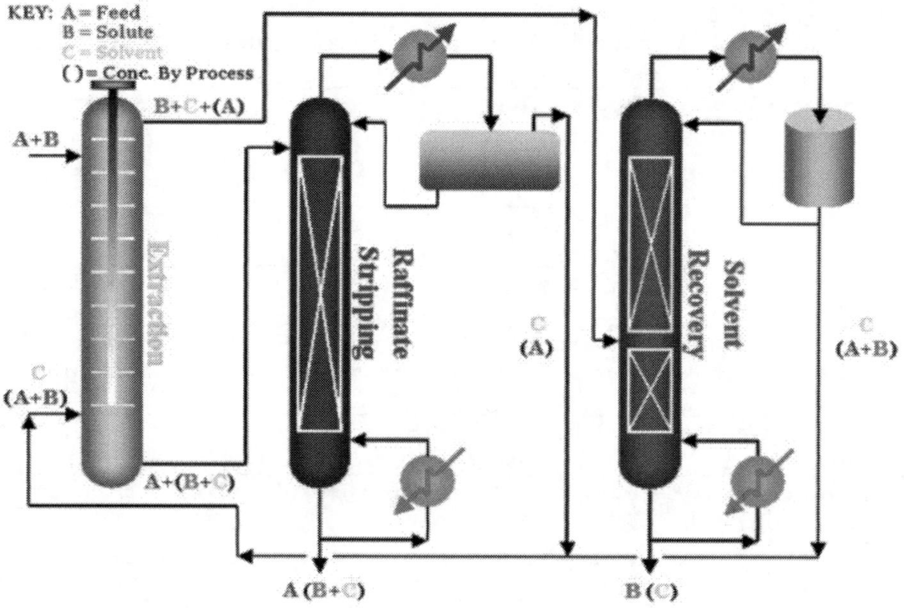

Figure 1.0 Typical Flowsheet of a Liquid Extraction B- Solute; C- Solvent

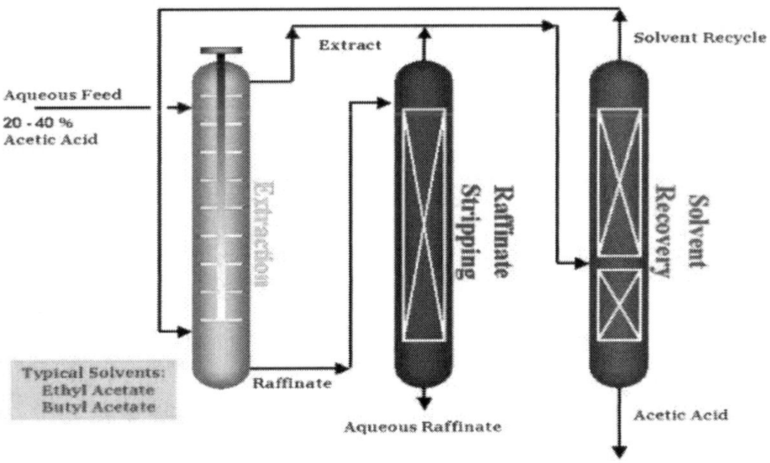

Figure 2.0 Recovery of tightly Hydrogen-Bonded Organics from Water; such as Formaldehyde, Formic acid and Acetic acid

A co-solvent can be used in addition to the solvent in order to extract more of the solute from the raffinate phase during continuous co-current multi-stage continuous liquid extraction operations. In Figure 1.0 is shown the schematic of the inclusion of the co-solvent.

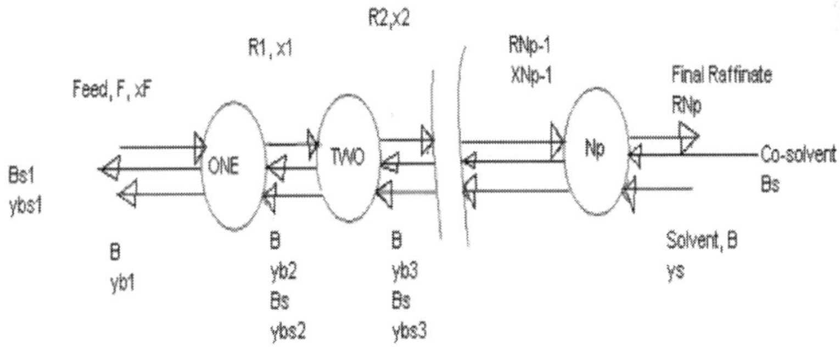

Figure 3.0 Use of Co-Solvent in Continuous Countercurrent Multi-Stage Liquid Extraction Operations

Fresh solvent and co-solvent is assumed without any solute in them. For the special case when the raffinate flow, $R_{NP} = R_{Np-1} = ... = R_2 = R_1 = F = R$ and when the two extract phases flow rate is constant in all the stages and is equal to B and B_s respectively. The equilibrium relationships between the distribution of solute between the extract and raffinate phases can be assumed to be linear. This is so the case when the solutions are dilute. Thus,

$$y_{Bk}^* = m_1 x_k \quad (1)$$

$$y_{Bsk}^* = m_2 x_k \quad (2)$$

A solute component mass balance on stage 1 can be written as;

$$x_0 R + y_{bs2} B_s + y_{b2} B = x_1 R_1 + y_{bs1} B_s + y_{b1} B \quad (3)$$

Dividing throughout the Eq. [3] by B and using the Eqs. [1,2] to reduce the x_1 and x_0 values in terms of y_{b1} and y_{b0} values;

$$R y_{b0}/B m_1 + y_{bs2} B_s/B + y_{b2} = R y_{b1}/B m_1 + y_{bs1} B_s/B + y_{b1} \quad (4)$$

The extraction factor, A_f can be defined as $A_f = R/m_1 B$. Further it can be seen that,

$$y_{bs2} = m_2 x_2 = (m_2/m_1)\, y_{b2}$$

$$A_f\, y_{b0} + y_{b2} B_s m_2/B m_1 + y_{b2} = A_f\, y_{b1} + y_{b1} B_s m_2/B m_1 + y_{b1} \quad (5)$$

A co-solvent factor can be defines as,

$$C_s = B_s m_2/B m_1$$

$$A_f\, y_{b0} + C_s\, y_{b2} + y_{b2} = A_f\, y_{b1} + C_s\, y_{b1} + y_{b1} \quad (6)$$

$$\text{Or} \quad (1 + C_s)\, y_{b2} = y_{b1}(1 + A_f + C_s) - A_f y_{b0} \quad (7)$$

$$y_{b2} = y_{b1}(A_f/(1+C_s) + 1) - A_f y_{b0}/(1+C_s) \quad (8)$$

In a similar fashion

$$y_3 = \; = y_{b1}\,(1 + A_f/(1+C_s) + (A_f/1+C_s)^2) - y_0(A_f/(1+C_s) + (A_f/1+C_s)^2)\,) \quad (9)$$

Thus,

$$y_{N+1} = y_1(1 + A_f/(1+C_s) + ... + .A_f^{Np}/(1 + C_s)^{Np} - y_0(A_f/(1+C_s + \; + ... + A_f^{Np})/(1 + C_s)^{Np} \quad (10)$$

$$(A_f/(1 + C_s) - 1)y_{Np+1} = y_1 ((A_f/(1 + C_s))^{Np+1} - 1) - y_0A_f/(1+C_s) ((A_f/(1 + C_s)^{Np} - 1) \quad (11)$$

Eq. [11] is applicable when $A_f \neq 1$. In a similar fashion the raffinate compositions can be derived;

$$x_{Np} = x_1((A_f/(1 + C_s)^{Np} - 1)/(A_f - 1) - x_0A_f((A_f^{Np-2} - 1)/(A_f - 1) \quad (12)$$

Sources, Further Reading and References

Belter, P. A., Cussler, E. L., and Hu, W.S., 1988, *Bioseparations*, Wiley, New York.

Catani, S. J., Vernon, N. M., Neiditch, D. S., Wiley, J. E. and Micinski, E, 2006, US Patent 7,049,435, *Extractive Methods for Purifying Sucralose,* Tate and Lyle, UK.

Cusler, E. L., 1997, *Diffusion : Mass Transfer in Fluid Systems,* Cambridge University Press, UK.

Inernet Website, Extraction Equipment, Koch Modular Process Systems, 2006, *Extraction Technology Group*, Paramus, NJ, USA

Kohr, W.J.,Shrader, V., and Johansson, C., 2005, US Patent 6802888, *High temperature heap bioleaching process,* Geobiotics, CO, USA.

Laddha, G. S., and Degaleesan, T. E., 1976, *Transport Phenomena in Liquid Extraction*, Tata-McGraw Hill, N.Delhi.

Lo, T. C., Baird, M. H. I., and Hanson, C., 1983, *Handbook of Solvent Extraction,* Wiley, New York.

Marsden, J.O., Brewer, R. E., Robertson, J. M., Hazen, W., Thompson. P. B., David, R., 2006, US Patent 7041152, *Method for processing elemental sulfur-bearing materials using high temperature pressure leaching*, Phelps Dodge Corp., AZ, USA

McCabe, W. L., Smith, J. C., and Harriott, P., 1993, Unit Operations of Chemical Engineering, Fifth Edition, McGraw Hill, New York.

Meinhold, J. F., Musto, J. A., Kramer, K. C. and Gottesman, M., (deceased), 1987, *Method for the decaffeination of roasted coffee extracts*, US patent 4,659,577, General Food Corp., White Plains, NY, USA.

Sharma, K. R., *Recycle of Uranium in Fission Products*, AIChE Spring National Meeting, New Orleans, LA, USA, March, 2002.

Sharma, K. R., 2005, *Damped Wave Transport and Relaxation,* Elsevier, Amsterdam.

Rao, G. and Sittig, M., 1997, *Dryden's Outlines of Chemical Technology – For the 21st Century,* East-West Press, N.Delhi.

Treybal, R. E., 1963, *Liquid Extraction,* McGraw Hill, New York.

Treybal, R. E., 1980, *Mass Transfer Operations,* Third Edition, McGraw Hill, New York.

Predicting Pressure Drop and Holdup in Katapak-Sp Using a Particle Model

Ricardo Macías-Salinas, ESIQIE-Departamento de Ingeniería Química, Instituto Politécnico Nacional, UPALM, Av. IPN s/n, Col. Lindavista, México, D.F., Mexico, Laura Rosas-Ortiz, Petroleum Engineering Department, Instituto Politécnico Nacional, UPALM, Av. IPN s/n, Col. Lindavista, Mexico, D.F., Mexico, and Elisabetta Brunazzi, Chemical Engineering Department, University of Pisa, Via Diotisalvi 2, Pisa, Italy.

Various experimental efforts have been reported in the literature in an attempt to understand the hydrodynamic behavior of a relatively new family of catalytic structured packings: KATAPAK-SP. The only approach so far used to represent the experimental results obtained for KATAPAK-SP 11 & 12 relies in the channel model. The purpose of this work was to verify the suitability of a particle model originally developed for random and structured packings (Stichlmair et al., Gas Sep. Purif., 3, 19, 1989) by extending its applicability to catalytic structured packings, particularly for the KATAPAK-SP family. In doing that, a more suitable liquid holdup correlation for the open channels was proposed. The question on how liquid splits into the open channels and catalyst bags still remains partially unanswered. In an attempt to solve this, a new calculation procedure has been devised here to reasonably determine the liquid share between the open channels and the catalyst bags. Model predictions in terms of pressure drop and liquid holdup are presented and discussed in this work for the case of KATAPAK-SP 11 & 12.

Hybrid Microwave Heating of Palladium for Helium Release

Paul R. Burket, Process Science & Engineering, Savannah River National Laboratory, Building 773-42A, Savannah River Site, Aiken, SC 29808 and R. Tom Walters, Defense Process Technology, Savannah River National Laboratory, Building 773-42A, Savannah River Site, Aiken, SC 29808.

A technique is under development that releases the helium content from a palladium-tritide storage bed in a way that allows quantitative mass spectral measurement. Currently, helium from tritium decay remains trapped after delivery of the stored tritium gas, so a complete characterization of the storage bed is not obtained from the delivered gases. SRNL has successfully identified the right materials of construction and process configuration to heat small samples of palladium storage bed material in a commercial "hybrid" microwave oven in a controlled manner to cause the desired effect of releasing the helium. After a few successful tests using zirconia as a surrogate storage bed material, the first test using representative (cold) palladium storage bed material was performed. A temperature of 1200 C was obtained in minutes and was controlled for over half an hour. Phase one of the project is completed. The successful implementation of this inexpensive microwave oven process will not only complete the mass balance correctly, but it will also provide a potential cost savings for the program over alternative methods for heating the spent palladium-tritide storage material.

Enzymatic synthesis of medium-chain partial glycerides in solvent-free media with up-flow reactor

Yun Fang*, Xue-yi Hu, Yong-mei Xia

School of Chemical and Material Engineering, Southern Yangtze University,1800 Lihu Avenue, Wuxi, Jiangsu 214122, China

Abstract: Enzymatic synthesis of medium-chain partial glyceride without solvent was investigated using an enzyme packed bed reactor (EPBR). In the up-flow EPBR, at 55 °C, with an initial water content in glycerol of 4% (w/w), glycerol: caprylic acid of 1.1:1 (mol/mol) and a space velocity of 2.5 min^{-1}, conversion of caprylic acid reached 90.2% in 4 h. The EPBR represented a good operating stability; conversion of the carbonic acid dropped only 6.5% for caprylic acid, and 4.5% for capric acid after 30 runs.

Keywords: Esterification; Lipase; Microaqueous; Glyceride; Solvent-free

1. Introduction

Medium-chain partial glycerides (partial glycerol caprates or caprylates or their mixture) are widely used functional additives in medicine, cosmetics and food, which has been manufactured by reacting extra glycerol with capric acid or capric/caprylic acid mixture at 150 °C catalyzed by an inorganic acid. Enzymatic synthesis of medium chain glycerides has been reported since last decade [1–7]. The most economical enzymatic synthesis technique is the reaction in solvent-free media; most of documented results were based on reactions employing large lipase amount, long reaction time and conducted in batch reactors [1–4], while EPBR has not been documented well. In this study, enzymatic synthesis of medium-chain partial glycerides in solvent-free media was conducted in a packed bed reactor. Dehydration protocol, mechanical stability of the immobilized lipase and flow style in the reaction was investigated with a commercial lipase CAL, respectively.

2. Experimental

2.1. Chemicals

Lipase from *Candida antarctica* (CAL) with olive oil hydrolysis activity 7000 U/g was a gift from Novo Nordisk Co. Capric acid containing 0.10% H_2O (*w/w*), caprylic acid containing 0.11% H_2O (*w/w*) and glycerol containing 1.14% H_2O (*w/w*) were from Shanghai Chemicals Co., China. Monocaprin and caprin were from Sigma. All other reagents are of analytical grade. Reactants were used as received unless otherwise stated.

* Correspondence to: Yun Fang, School of Chemical and Material Engineering, Southern Yangtze University,1800 Lihu Avenue, Wuxi, Jiangsu 214122, China. E-mail: yunfang@126.com

2.2. Analysis

Water content was determined with Karl Fischer titration. Conversion of carbonic acid was defined as the percentage of consumed carbonic acid. After removing unreacted glycerol from the reaction mixture, the composition of glycerides was determined by HPLC, employing a C_{18} column, at column temperature of 40 °C, methanol (1.0ml/min), and UV detector detecting at 215 nm.

2.3. Enzymatic esterification in BR

For comparison, the enzymatic esterification was also conducted at 55 ± 0.5°C in a 250 mL opened vessel, with an initial water content in glycerol of 4% (w/w), glycerol: carbonic acid of 1.1:1 (mol/mol), enzyme loading of 100 U /g carbonic acid, 400 r/ min.

2.4 Enzymatic esterification in EPBR

Two different processes were tested. The reaction system (Scheme 1) was consisted of a water-jacketed EPBR, a peristaltic pump and an opened water-jacketed vessels equipped (one vessel in loop A, two vessels in loop B) with a mechanical stirrer. The EPBR was a glass column（8 mm in inner diameter, 12 cm in height） filled with 2 g CAL, the catalyst bed was 10 cm in height. In the first vessel the reactants (1) were well mixed, pre-heated to 55 ± 0.5 °C, and then pumped (2,3) through EPBR (pre-warmed to 55 ± 0.5°C). In loop A, the reaction mixture from EPBR (4) was delivered back to the vessel to dehydrate and mixed with the fresh raw materials (5) before the second reaction cycle; While in loop B, the reaction mixture from EPBR (4) was delivered to the second vessel to dehydrate via vaporization (5) until the first vessel emptied the raw materials, and then the mixture went through the EPBR (6,7) to the first vessel for further dehydrating (8,9) before the second reaction cycle. That is, in loop B, an extra vessel was employed to further dehydrate. The reactor was also opened to enhance dehydration. The flow rate was monitored by a flow meter. Unless otherwise stated, the reaction was conducted at 55 ± 0.5°C, with an initial water content in glycerol of 4% (w/w), 0.5mol carbonic acid, glycerol: carbonic acid of 1.1:1 (mol/mol) and a space velocity of 2.5 min^{-1}.

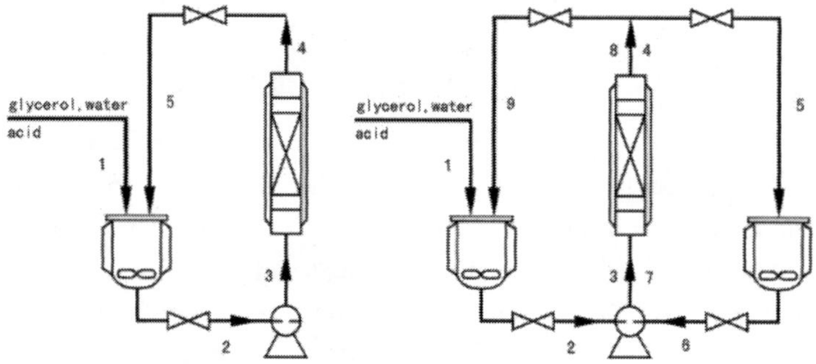

A B

Scheme 1 Schematic reaction EPBR loops

3. Results and discussion

3.1 Comparison of reactions conducted in EBR and EPBR

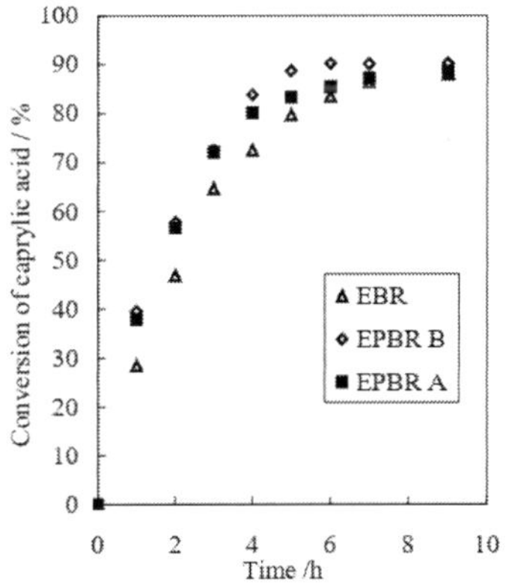

Reaction conditions: 55 °C, initial water content in glycerol was 4% (w/w), 0.55 mol glycerol, 0.5 mol caprylic acid, space velocity was 2.5 min^{-1}; enzyme loading was 2g in EPBR and 100U/g acid in EBR.

Figure 1 Time courses of reactions conducted in different reactors

Figure 1 shows that EPBR was more efficient than EBR, and loop B (EPBR B) resulted in a noticeable improvement than loop A (EPBR A) did due to the further dehydration. In the subsequent study, loop B was employed thereafter.

3.2 Flow style of the reactor

Table1 Effect of flow style on the esterification

Flow style	Conversion of caprylic acid / %	Conversion of capric acid / %
Up-flow	90.10	93.25
Down-flow	90.35	92.56

Reaction conditions: 55 °C, initial water content in glycerol was 4% (w/w), 0.55 mol glycerol, 0.5 mol caprylic acid, space velocity was 2.5 min^{-1}, 2g enzyme,6hr

Although the flow style didn't change the conversion dramatically, and downstream flow could avoid back-mixing, but it pressed the catalyst bed especially after several runs. Therefore in the following investigation, up-flow style was used throughout.

3.3 Effect of space velocity on the esterification

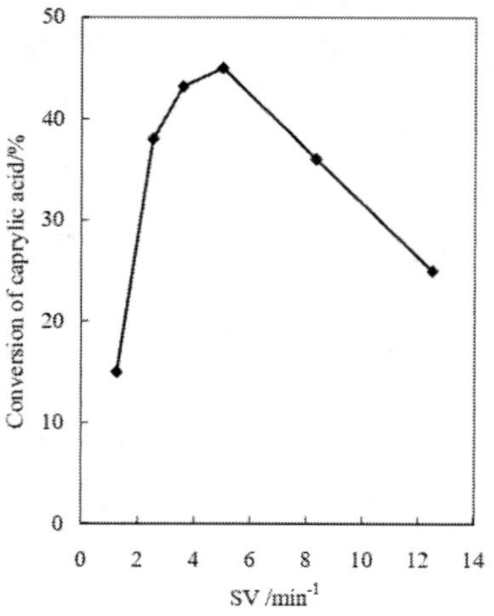

Reaction conditions: 55 °C, initial water content in glycerol was 4% (w/w), 0.55 mol glycerol, 0.5 mol caprylic acid, 2g enzyme, 1.5hr

Figure 2 Effect of space velocity on the esterification

2.5-5 min^{-1} was found to be the suitable space velocity range from figure 2. To maintain the operation flexibility, 3.5 min^{-1} is the optimum space velocity. However, after 10 runs, there was no significant difference between the conversions obtained with 2.5 min^{-1} and 3.5 min^{-1}, therefore, 2.5 min^{-1} has been selected.

3.4 Dehydration

It is a well-known fact that suitable water content is essential in enzymatic reaction. Molecular sieves, vacuum and plain evaporation are commonly used dehydration protocols in esterifation. Obviously, the last approach is the cheapest and fit to the case where essential water content for the reaction is relatively flexible, for example, the case of the enzymatic glycerylation [6]. However, to compare the dehydrate approaches, a dewatering column (same size and loading capacity as the EPBR) filled with 4A molecular sieves was placed in the outlet of the EPBR to be the alternative of the second opened vessel. As one can see from Table 2, the molecular sieves column didn't help obviously although the dehydrating capacity was enough. Whereas, the extra molecular sieves column caused more flow resistance.

Table2 Effect of molecular sieve column on the conversion of caprylic acid

Dehydration protocol	Conversion of caprylic acid / %
Opened vessel	89.5
Dewatering column filled with 4A molecular sieves	90.5

Reaction conditions: 55 °C, initial water content in glycerol was 4% (w/w), space velocity was 2.5 min^{-1}, 0.55 mol glycerol, 0.5 mol caprylic acid, 2g enzyme, 5.5hr

3.4 Enzyme stability

Table 3 The enzyme activities in different locations of the catalyst bed

Enzyme source	*Conversion of*	*Conversion of*

	caprylic acid/%	capric acid/%
Fresh enzyme	89.8	92.4
Top (1 cm)	85.7	90.5
Middle (5 cm)	81.5	86.2
Bottom (10 cm)	78.4	83.4

Reaction conditions: 55 °C, initial water content in glycerol was 4% (w/w), space velocity was 2.5 min^{-1}, 0.55 mol glycerol, 0.5 mol caprylic acid, 2g enzyme, 6hr per run, 30runs

After running the up-flow reactor for 30 runs, the enzyme was taken out from different position in the catalyst bed to exam the enzyme's stability. Table 3 shows that the enzyme was deactivated in different degree depending on its source location in the reactor. So, if we replace the enzyme from the bottom of the bed once by once in every month, it is possible to make a constant active reactor.

3.5 Operation stability of the system

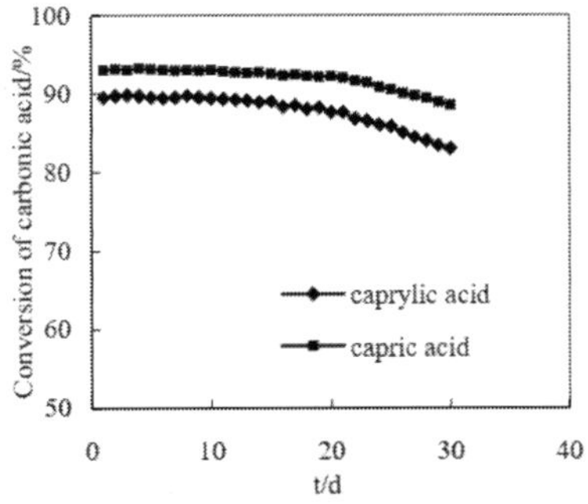

Figure 3 Time course in the up-flow EPBR in 30days

The EPBR represented a good operating stability; conversion of the carbonic acid dropped only 6.5% for caprylic acid, and 4.5% for capric acid after 30 runs.

4 Conclusion

Enzymatic synthesis of medium-chain partial glyceride without solvent was achieved using an up-flow enzyme packed bed reactor (EPBR), in which the conversion of caprylic acid reached 90.2% in 4 h. Dehydration with a molecular sieves column or via vaporization by an opened vessel yielded similar results. 2.5-5 min^{-1} was found the suitable space velocity range. The flow style didn't change the conversion dramatically. The enzyme was deactivated in different degree depending on its source location in the reactor. The EPBR represented a good operating stability; conversion of the carbonic acid dropped only 6.5% for caprylic acid, and 4.5% for capric acid after 30 runs.

Acknowledgments

This work was supported by Program for New Century Excellent Talents in University by Ministry of Education of China, NSFC project (20476038), 863 project of MOST of China and Jiangsu Provincial NSF of China (BK2006021). We thank Mr. Hu Jian and Miss Xu Qing in Novo Nordisk Co. for generously providing lipase. We also thank Dr. Andrew Tadd at University of Michigan for his helpful revision.

References

1 Nandi S, Gangopadhyay S, Ghosh S. Production of medium chain glycerides from coconut and palm kernel fatty acid distillates by lipase-catalyzed reactions. *Enzyme And Microbial Technology* 2005, 36 (5-6): 725-728

2 Saxena, RK; Isar, J; Saran, S; Kaushik, R; Davidson, WS. Efficient microwave-assisted hydrolysis of triolein and synthesis of bioester, bio-surfactant and glycerides using Aspergillus carneus lipase. *Current Science,* 2005, 89 (6): 1000-1003

3 Wong, WC; Basri, M; Razak, CNA; Salleh, AB. Synthesis of medium-chain glycerides using lipase from Candida rugosa. *Journal Of The American Oil Chemists Society*, 2000, 77 (1): 85-88

4 Yesiloglu, Y; Kilic, I. Lipase-catalyzed esterification of glycerol and oleic acid. *Journal of the American Oil Chemists Society*, 2004, 81 (3): 281-284

5 Pereira, CCB; Da Silva, MAP; Langone, MAP. Enzymatic synthesis of monolaurin. *Applied Biochemistry and Biotechnology*, 2004,113: 433-445

6 Xia Y M, Fang Y, Zhang K C, Shi G Y, Brown J J. Enzymatic Synthesis of Partial Glycerol Caprate in Solvent-Free Media. *J. of Molecular Catalysis B: Enzymatic*, 2003, 23(1): 3-8

7 Da Silva, MAM; Medeiros, VC; Langone, MAP; Freire, DMG. Synthesis of monocaprin catalyzed by lipase. *Applied Biochemistry and Biotechnology*, 2003, 105: 757-768

8 Marta A P L, Melissa E D A. Enzymatic synthesis of medium chain mono-glycerides in a solvent-free system. *Applied Biochemistry and Biotechnology*, 2002, 98-100: 987-996

9 Langone, MAP; Sant'Anna, GL. Enzymatic synthesis of medium-chain triglycerides in a solvent-free system. *Applied Biochemistry and Biotechnology*, 1999, 77-9: 759-770

10 Shanmugam, VM; Sasikumar, PG; Rao, NM. Characterisation of acylglycerol synthesis in aqueous foams by lipases. *Journal of Chemical Technology and Biotechnology*, 1999, 74 (6): 515-518

11 Goto, M; Hatanaka, C; Haraguchi, T. Glyceride synthesis using lipase immobilized in N-polyisopropylacrylamide gel. *Kagaku Kogaku Ronbunshu*, 1998, 24 (6): 923-927

12 Kim S M, Rhee J S. Production of medium-chain glycerides by immobilized lipase in a solvent-free system. *Journal of the American Oil Chemists Society,* 1991, 68: 499-503

Improving Mercury Removal Efficiency of a Plasma Enhanced Electrostatic Precipotator

Sreerupa Basu, Chemical and Biomolecular Engineering, Ohio University, Athens, OH 45701

Chlorine atom oxidizes elemental mercury to mostly soluble species. Addition of chlorine gas into a plasma enhanced electrostatic precipitator, used to clean flue gas from coal combustion, will thus increase the mercury removal efficiency in the precipitator. Determination of the chlorine atom concentration, formed inside the chamber, is a key to evaluate this efficiency. A series of experiments are performed to dissociate the chlorine gas in a corona-discharge field formed inside a 15 cm by 3 cm flow pyrex tube at P=1atm, and the chlorine atoms formed are measured by reacting them with butane. The reaction products are quantified using a FTIR. Determination of the chlorine atom concentration in the plasma region near the charging electrode, as a function of voltage supplied, quantity of chlorine injected and the residence time allowed to the chlorine gas between the electrodes will be the first step in optimizing the amount of chlorine reagent gas needed to be added to a precipitator to obtain enhanced mercury removal efficiency. Residence time in the order of milliseconds is experimented to track the behavior of the chlorine atoms in chlorine plasma. This aims at understanding if the 1 minute residence time of industrial precipitators is enough or too much to optimally charge elemental mercury in flue gas stream to obtain a more than 99.9% mercury removal efficiency target for a plasma enhanced electrostatic precipitator.

Fast and non-invasive measurement of cross-link density in polymers by TD-NMR

Supriyo Ghosh, Z Harry Xie, Bruker Optics Inc., The Woodlands, TX

Abstract

Mechanical and thermal properties of the cured polymer products are strong functions of their cross-link density (CLD). A continuous monitoring of the curing process is done through measurement of CLD during production. Swelling and rheological measurements are the two most commonly used routine CLD measurement methods in the polymer industry. The swelling experiment is time consuming, poorly repeatable, involves handling of organic solvents, and has ample opportunity to introduce human error. The rheological measurements, although mechanized, have poor repeatability and not fast. A fast method of measuring CLD in polymers would enable better process monitoring and quality control in cross-linked/cured polymer products. A fast, non-invasive, and automated TD-NMR (time-domain nuclear magnetic resonance) measurement of polymer CLD can be done using a small bench-top instrument without any sample preparation. In TD-NMR, the signal is obtained from the hydrogen atoms present in the polymer molecules and it varies with their mobility. When cross-linking is introduced in a polymer, TD-NMR looks directly at the molecular level and measures the mobility of the polymer strands, which in turn provides a measure of the CLD. The basics of TD-NMR, as applies to quantitative study of polymers, will be elaborated in the presentation. Through quantitative analyses of the NMR signals from various polymers, it is shown to correlate very well with the CLD of the sample. Experimental data from the TD-NMR measurement of PE-foam samples will be shown as an example. This easy TD-NMR measurement of CLD seems to have promise in process monitoring and quality control of cross-linking in polymer industry.

Introduction

Polymers are produced for a wide variety of end-use. One of the most important functional properties of polymers is their mechanical strength or rheological behavior. The network structure in the polymer matrix needs to be engineered properly in order to attain a certain rheological property in them. The cross-link density is one of the crucial network structural features in this regard. Hence, a quantitative measurement of the cross-link density of polymers would have large contribution towards predictive modeling of thermo-mechanical behavior of a specific polymer sample. This quantitative assessment of polymer samples can not only help us predict the polymer properties but also a regular monitoring of the cross-linking in a curing process can help realized a feedback control system, improving the economy of the process. During the end-use of a polymer material, it experiences various stresses e.g. mechanical, radiation, thermal, chemical, etc. These environmental factors gradually modify the network structure; the cross-link density changes over time. A monitoring of the kinetics of cross-link density change under those environmental conditions can help predict the probability of failure of the polymeric materials and design better polymers.

Although quantitative assessment of the extent of cross-linking is very important in polymer industry, there is no easy and fast routine method for it. Swelling and rheological methods are the current industry standards for cross-link density monitoring. Swelling involves use of solvent chemicals, the measurement procedure is very lengthy and has ample opportunity for introduction of human error. The rheological methods are time-consuming, and the repeatability of the cross-link density prediction is poor. None of these methods can be automated.

Nuclear magnetic resonance (NMR) is one of the advanced technologies for non-invasive monitoring of material properties. NMR can monitor the molecular motion in a sample and quantify them. This capability of NMR is utilized in the measurement of cross-link density in polymers. The nuclei of certain atoms e.g. hydrogen have electric charge as well as spin, as a result of which they have magnetic moments. These magnetic moments remain randomly oriented in samples, and thus the sample has no detectable magnetization vector. During an NMR measurement, as the sample is placed in a magnetic field, the nuclei of specific atoms (e.g. hydrogen nuclei) align with the external magnetic field, producing a detectable magnetization. A single radiofrequency (R.F.) pulse or a combination thereof, which is/are in resonance with the nuclear spin frequency, is then applied to the sample to flip the ensemble average of the magnetization by a specific angle. This rotating magnetization generates electrical signal in the R.F. antenna, and the analog signal is amplified, sampled and stored as the discrete time-domain NMR signal for further analysis.

Cross-linked polymers can be envisioned as a network of two types of chains: chain segments restricted at both ends by two different cross-linking points, and chain segments that are restricted only at one end. The hydrogen nuclei or protons present in these segments interact with each other. The nuclear magnetization that is observed in an NMR experiment, relaxes due to this interaction between two adjacent protons (dipolar interaction) through a transverse relaxation pathway. If the sample is heated and brought to a temperature that is well above its glass transition temperature (T_g), then the thermal motion of the free chains average out the dipolar interaction. Two types of motions are observed in the polymer network when it is well above the T_g: a slow motion caused by the reptation and tube-renewal characterized by a correlation time τ_s of about 10^{-3} second, and a fast local motion with a correlation time τ_f of about 10^{-9} second. The slow motion is almost isotropic in the aforesaid condition and cause motional narrowing in the NMR signal. More restrictions the chains have due higher amount of cross-linking, more anisotropic the fast local motion becomes. Because of this anisotropy all of the dipolar couplings do not get averaged out. This is manifested in the Gaussian shape contribution in the relaxation curve of time-domain NMR data, as described by equation 1.[1]

$$M(t) = A\exp\left\{\frac{-t}{T_2(\tau_f)} - qM_2\tau_s^2\left[\exp\left(\frac{-t}{\tau_s}\right) + \frac{t}{\tau_s} - 1\right]\right\} + B\exp\left(\frac{-t}{T_2(\tau_f)}\right) \quad (1)$$

where M(t) is transverse magnetization, A is the mass fraction of the chains restricted at both ends, B is the mass fraction of the chains restricted only at one end, qM_2 is the residual contribution of the second moment of the rigid lattice(M_2), T_2 is transverse relaxation rate constant. Thus, the ratio of the second moment due to the residual dipolar coupling and the second moment of the rigid lattice gives a quantitative estimate of the extent of cross-linking present in a polymer sample. This value is often referred to as the q-value. Sometimes, if there is an extensive amount of physical entanglement present in a polymer network, it contributes partially to the residual dipolar coupling. This value is usually quite low when the sample is tempered well above the T_g. Hence, to get the true value of chemical crosslink density, the measured q-value of the sample needs to be corrected with the q-value obtained by analyzing the same polymer sample without any chemical cross-linking. The chemical crosslink density of a polymer can be calculated from the measured q-value if the following information are known: molecular mass of one monomer unit per number of backbone bonds in one unit, number of backbone bonds in one Kuhn-segment.

The NMR instruments usually used in the research laboratories operate at a very strong magnetic field using superconducting magnet, maintained using cryogenic systems. They need large space for installation, often expensive, need regular maintenance and technical personnel's with extensive NMR experience and expertise. These instruments can provide the researcher an enormous wealth of information related to material characterization. But it is not essential to use this high-resolution NMR in order to get the crosslink density information from a polymer sample. The time-domain NMR data that is required to quantify

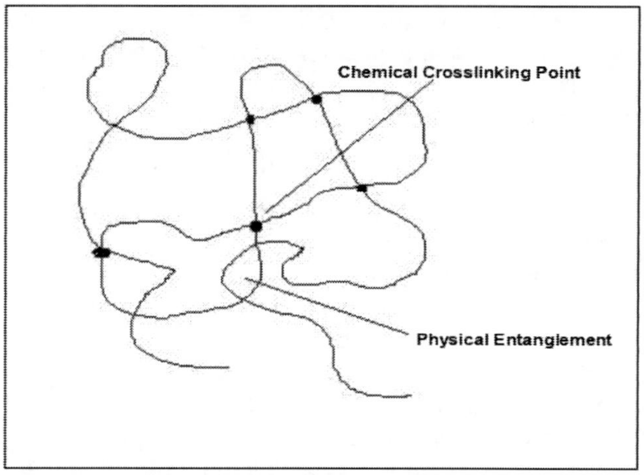

Figure **1**: Schematic diagram of polymer network

the extent of cross-linking, has been successfully obtained from small, bench-top TD-NMR instrument like Bruker minispec. This instrument uses permanent magnet and hence needs no regular maintenance of cryogenic system. It uses comparable or most often weaker magnetic field than a clinical MRI system, and hence can be used in industrial production area for close process monitoring. Since the low-field time domain NMR instrument does not need either any instrumental adjustment for each sample or any sample preparation, the whole process of crosslink density measurement and data management is done automatically with the help of a robotic system. This enables a polymer industry to have a high-throughput, consistent crosslink density measurement procedure without any human intervention or error.

Materials and Methods

A Bruker minispec mq20 TD-NMR instrument (Bruker Optics Inc, Billerica, MA) with a high temperature probe was used for all the measurements described in this publication. The TD-NMR instrument was tuned at 20MHz frequency for hydrogen. The sample cavity inside the NMR probe was insulated from the magnet, and the sample was heated continuously using hot air. The flow rate and the temperature of the heated air were maintained constant using a PID controller.

Three different samples of polyethylene (PE) foams, manufactured in a pilot plant with different amount of cross-linker added to them, were measured using the instrument. Three more samples with high, medium and low levels of cross-linking as well as a group of randomly picked samples from a production line were measured using the same technique. A few rubber samples with different amount of cross-linker added were also measured.

After the sample are put inside the tubes and loaded into the rack, the robotic system took the samples to the probe based on the priority level defined by the user. The sample stayed inside the probe for an optimized pre-heating time. The preheating time widely varies among samples. It depends on the thermal conductivity and specific heat of the sample being measured. A spin-echo NMR measurement started at the end of the pre-heating time period. The spin-echo measurement progressed as per the user defined array of echo-time values.

Figure **2**. Automated TD-NMR crosslink density analyzer

The time-domain NMR data looked like figure 3. As soon as the measurement was over, the time-domain data was processed by the minispec software, which controlled the whole instrument, to output the second moment or the q-value.

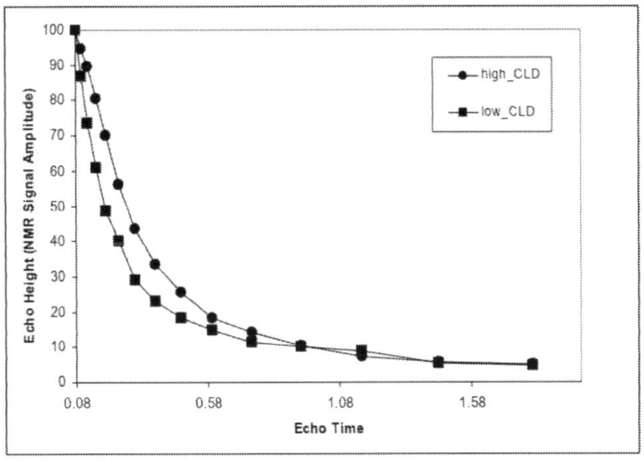

Figure **3**. TD-NMR data recorded by Bruker minispec for q-value calculation

Results and Discussion

Three PE foam samples with high medium and low cross-linking were measured at 90°C temperature after a preheating time of 25 minutes. The q-values, as shown in figure 4, were in line with the extent of cross-linking in them. Nine samples were drawn randomly from the production line on two different days. These PE foam samples were also measured for q-values using Bruker minispec. The variation in cross-linking achieved in a curing process can thus be monitored closely as shown in figure 5.

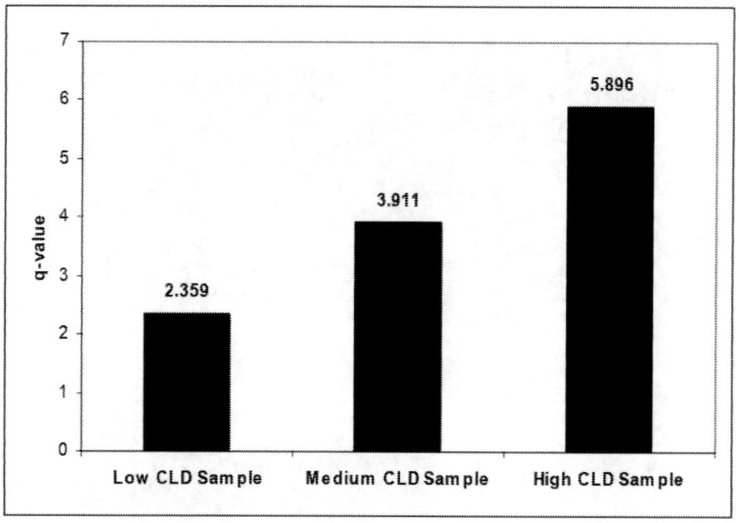

Figure **4**. q-values of polymers measured using Bruker minispec TD-NMR

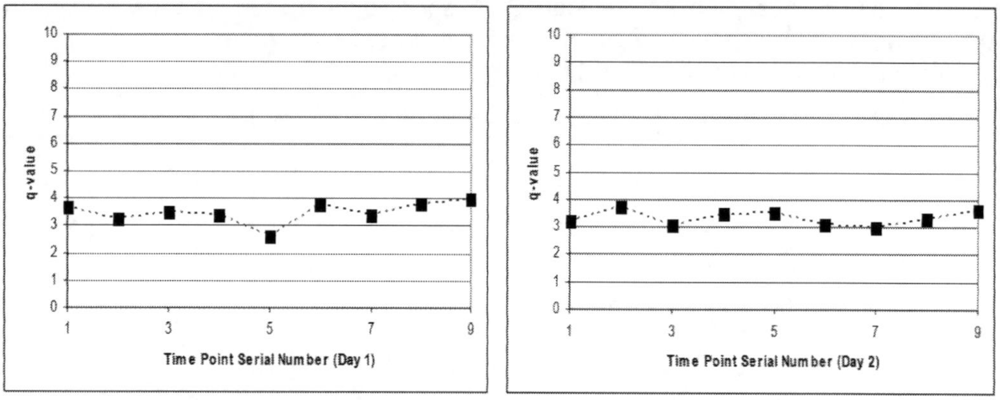

Figure **5**. Easy and fast curing process monitoring using Bruker minispec

In order to check the repeatability of the TD-NMR crosslink density measurement procedure, one sample was measured five times followed by identical preheating procedure. The repeatability was measured calculating the standard deviation of these consecutive measurements. The standard deviation was found to be 0.034, which corresponds to about 0.05phr difference in cross-linker concentration.

Table **1**. Bruker minispec TD-NMR repeatability of crosslink density measurement

Replication No.	q-value
1	2.359
2	2.416
3	2.336
4	2.396
5	2.406
Std. Dev.	0.034

To validate the prediction of extent of cross-linking, three PE foam samples prepared in pilot plant using three different levels of cross-linker added to it were measured using the same measurement sequence. The level of cross-linker added was observed to have a very high correlation with the q-value calculated by Bruker minispec, as shown in figure 6.

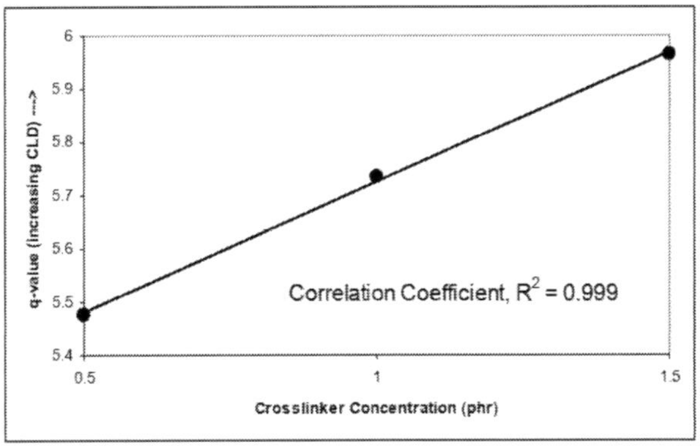

Figure **6**. Correlation between added cross-linker amount and q-value

Acknowledgements

We would like to acknowledge the assistance of Dr Gisela Guthausen of Bruker Germany in terms of experimentation method development and programming for data analysis.

References

1. Menge H, Hotopf S, Schneider H. Characterization of elastomeric polymer networks by transversal NMR relaxation. Kautschuk und Gummi Kunststoffe. 1997;50:268-73.

AUTHOR INDEX

American Institute of Chemical Engineers
3 Park Avenue
New York, NY 10016-5991

ISBN: 978-1-60423-361-2